D0847007

THE
ELEMENTS
BIBLE

A FIREFLY BOOK

Published by Firefly Books Ltd. 2018

Copyright © 2018 Octopus Publishing Group Ltd
Text Copyright © 2018 Rebecca Mileham

All rights reserved. No part of this publication may be reproduced, stored in a retrieval system, or transmitted in any form or by any means, electronic, mechanical, photocopying, recording or otherwise, without the prior written permission of the Publisher.

First printing

Publisher Cataloging-in-Publication Data (U.S.)

Library of Congress Control Number: 2018938722

Library and Archives Canada Cataloguing in Publication

Mileham, Rebecca, author
    The elements bible : the definitive guide to 350 years of scientific discovery / Rebecca Mileham.
Includes index.
ISBN 978-0-228-10113-0 (softcover)
    1. Periodic table of the elements. I. Title.
QD467.M55 2018   546'.8   C2018-901849-6

Published in the United States by
Firefly Books (U.S.) Inc.
P.O. Box 1338, Ellicott Station
Buffalo, New York 14205

Published in Canada by
Firefly Books Ltd.
50 Staples Avenue, Unit 1
Richmond Hill, Ontario L4B 0A7

Printed in China

First published in by Cassell,
a division of Octopus Publishing
Group Ltd
Carmelite House
50 Victoria Embankment
London EC4Y 0DZ
www.octopusbooks.co.uk

Edited and designed by Whitefox

Rebecca Mileham asserts the
moral right to be identified as the
author of this work.

Publishing Director Trevor Davies
Production Controller Katie Jarvis

THE ELEMENTS BIBLE

REBECCA MILEHAM

FIREFLY BOOKS

CONTENTS

INTRODUCTION

WHERE DID ALL THE ELEMENTS ON EARTH COME FROM?

WHICH METALS MAKE THE BEST COLOURS FOR FIREWORKS?

HOW CAN IODINE SAVE YOUR LIFE?

WILL WE EVER DISCOVER AN ELEMENT BEYOND NUMBER 118?

A journey through the periodic table takes you all around the world, through the inner layers of the planet, and out across landscapes of life-giving and lethal substances. You splash in oceans of the incredibly rare, the utterly ubiquitous and the downright odd. You meet metals and minerals, scale heights that glow and glitter, and take in views of stunning colour and variety. You hear, see and feel the stuff of life on Earth. You float through our precious atmosphere and head off into the stars. It's an amazing trip.

In this exploration of the elements, I begin with tales of the periodic table itself, and then spend time in the story of hydrogen, element number 1 and the most common in the universe. We finish with oganesson, element number 118, a fleetingly created and unstable heavyweight substance. The whole range and number of elements in between, the solids, liquids and gases of increasing weight, we meet within groups that share features – from the fiery alkali metals on the left, to the unflappable noble gases on the right.

▲ *The Danakil Depression in Ethiopia has a dazzling elemental landscape of lava lakes, hot springs and colourful deposits of sulfur and potassium salts.*

▶ *Iodine is an effective antiseptic.*

FINDING THE ELEMENTS

We gain a glimpse of the early chemists' experience: a revelation of the existence of many individual substances, and their intriguing characteristics. With a few exceptions – including oxygen and nitrogen in the air, plus gold, silver and sulfur – the elements aren't just lying about in handy chunks awaiting collection. Many individual elements proved tantalizingly tricky to track down – and, over the last few decades, finding new elementary substances has meant creating them for ourselves.

Cracking all the elements has therefore involved thousands of people, over hundreds of years, across dozens of countries. In laboratories around the world, scientists have invented new experimental techniques and apparatus to pull things apart, measure their properties and identify their components. Along the way, creative thinkers have come up with new uses and applications for almost every substance and its compounds.

▼ Solids, liquids and gases...the elements have a vast range of characteristics.

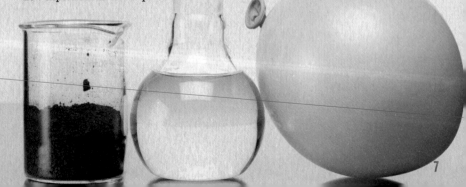

ATOMIC STRUCTURE

3

A quest for simplicity in complexity has often served scientists very well. The underlying structure of the atom, with its nucleus of protons and neutrons, and outer layers of electrons, proved to be the key to understanding the elements. It transpired that the secret behind an element's position in the periodic table is the number of protons in its nucleus. The reason for its chemistry and family features turned out to be because of how its electrons are arranged.

The discovery of elements is a drama that may never end. At the time of writing, the periodic table has no gaps; all the elements are confirmed and have received official names. However, more new elements may be waiting in the wings. A few weeks ago, I rediscovered a periodic table model given to me by my former chemistry teacher, Ted Lister, who was also the source of some of my first

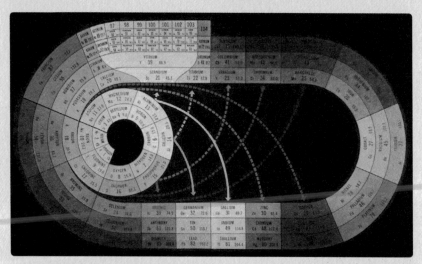

▲ Time *magazine published this spiral-shaped periodic table in 1949, beginning with hydrogen at the centre and using colours to link elements with similar properties.*

PERIODIC DIVIDE

Noble Gasses

Alkali Metals

Superactinides

Lanthanides and Actinides

Transition Metals

▲ *Otto Theodor Benfey published his periodic table, dubbed the Periodic Snail, in 1964. The elements again proceed in a spiral but the structure allows for the transition metals, as well as substances beyond the actinides.*

opportunities in science writing. The fact that the model stops at element 109, and that some of the controversial names of the heavy elements have altered and swapped around since then, reinforces the changing and developing nature of chemistry.

Elements 119 and onward could be synthesized in the future, and would presumably begin an eighth row, or period, of the table. Their electrons would be in new and different arrangements. It's possible that the inner electrons would be travelling near the speed of light, moving so fast that they gain mass according to Einstein's theory of special relativity. They may behave oddly.

LIFE-CHANGING ELEMENTS

Every current element has its own profile in the book – a fact-file of characteristics and sources. What I have found so intriguing is how these details, and the stories you find in each chapter, answer the question of how each element has changed human lives.

From the wisdom of ancient peoples who knew of magnetism, platinum-working and steelmaking, to the latest discoveries in medicine, engineering and space travel, I have let the stories range as widely as possible. Research led me into many areas of contemporary science as well as history, and I also enjoyed returning to a couple of textbooks from my student days in the physics department at Imperial College London.

▼ *Stories of the elements take us from magnetism to medicine, steelmaking to space travel.*

▲ *Geophysicist Mary Fowler is Master of Darwin College, Cambridge, and great grand-daughter of Ernest Rutherford.*

THE PIONEERS

I tried to include lots of the characters whose personal stories illuminated corners of the periodic table in memorable ways. Among dozens of brilliant people, some are featured as Periodic Pioneers, as my daughter Natalie named them. Dmitri Mendeleev is one, who published the first functional periodic table in 1869. Ernest Rutherford is another, mentioned above. Marie Curie, the pioneer of radioactivity, follows – the only person to win Nobel Prizes in both physics and chemistry. Then there is Lise Meitner, who never won a Nobel Prize

One such book was *Elementary Particles* by David Griffiths, the other *The Solid Earth* by Mary Fowler. This second book helped me dig deeper into the story of Inge Lehmann, a Danish seismologist whose work showed that the Earth's inner core is a solid sphere of iron. However, in writing an earlier chapter of the book, I had also been delighted to discover that Fowler – a distinguished geophysicist – is the great grand-daughter of Ernest Rutherford, whose scientific work and laboratory leadership led to the understanding of the structure of the atom.

▲ *Pioneer of the periodic table, Ernest Rutherford.*

but had an element named in her honour. The discoverer of heavy elements, Glenn Seaborg, comes next – another person whose name appears among the elements – and then Kosuke Morita, who was the first Asian scientist to name an element.

Despite the diversity of these pioneers, I was startled to notice in my research a tendency for people involved in the history of the elements to have the names William and Henry. In no particular order, the book's tally of Henrys includes: Henry Cavendish, Heinrich Geissler, Martin Heinrich Klaproth, Johann Heinrich Schulze, Henry Roscoe, Henry Bessemer, Heinrich Rose, Henry Moseley and Henri Moissan. When it came to Williams, I included: Robert Wilhelm Bunsen, Carl Wilhelm Scheele, August Wilhelm Hofmann, William Brownrigg, William Gregor, William Hillebrand, William Crookes, William Withering, William Hyde Wollaston, William Merriam Burton, Wilhelm Roentgen, William Shockley, William Lawrence Bragg, Wilhelm Blitz, William Ramsay and William Buehler.

Some people ticked both the boxes: William Henry Perkin, William Henry Fox Talbot and William Henry Bragg.

▲ *William Henry Perkin made great discoveries – but it wasn't his names that made him so gifted.*

▲ *Kosuke Morita discovered element 113 and named it nihonium.*

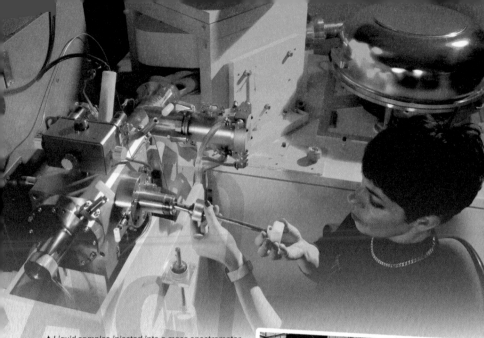

▲ Liquid samples injected into a mass spectrometer are bombarded with electrons to create ions, which then travel through electric and magnetic fields to reach a detector.

All these people made groundbreaking discoveries – but surely we wouldn't conclude it was because their names gave them intrinsic scientific prowess. More likely that, over the last few centuries, if you were called William or Henry you were more likely to have the cultural access and science capital to pursue scientific research at all.

But, ultimately, I found all the elements of a good story in the story of the elements. There were coincidences

▲ A mass spectrometer in use in the UK in 1949.

and mishaps, victories and tragedies, leaps of inspiration and dogged years of experimentation. I was glad to encounter multiple examples of scientists from contrasting places and walks of life who had the brains, the insight, the tenacity and wit to advance our understanding. In another hundred years, perhaps the range of names in scientific research will be considerably wider again than I have been able to make it here.

THE WEIGHT OF ATOMS

We know that all of the natural and artificial things around us are made from 118 different substances. But as atoms are so tiny, how did scientists measure the elements to determine their atomic weights?

In 1918, Canadian-American physicist Arthur Jeffrey Dempster built the first modern mass spectrometer. Inside, he could bombard samples of an unknown substance with electrons, causing ions – charged particles – to break off and fly along a tube through a magnetic field. Since ions of different mass would be deflected by characteristic amounts, Dempster could then identify the substance.

MASS SPECTROMETER

If an atom or molecule has an electric charge, its path will curve as it passes through a magnetic field. So, first of all, the elements in a sample are ionized – turned into charged particles – by knocking off electrons.

Then the ion stream is accelerated through a magnetic field. The deflection in the field depends on the ions' mass. The heavier the ions in the stream, the less their path curves. In addition, the amount of charge affects the deflection.

Ions are detected and identified at the other end of the tube. By varying the magnetic field you can detect ion streams of different substances.

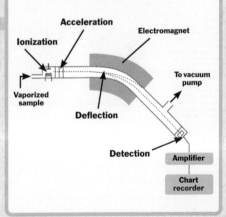

▲ As well as calculating a precise atomic weight for oxygen, Edward Morley worked with Albert Michelson on a famous experiment trying to detect the luminiferous ether in which scientists thought light travelled.

In 1919, in the UK, Francis William Aston also built a mass spectrometer. He used it to show that lots of common elements, including neon, chlorine and bromine, lithium, potassium, calcium and tin, existed as atoms of different weight, called isotopes. It meant some of their atoms had differing numbers of neutrons – and it made sense of some puzzling experimental results. Aston was awarded the Nobel Prize for Chemistry in 1922 for his find.

But there had been many years of research prior to the mass spectrometer. Almost every 19th-century chemist, for example, spent time trying to determine atomic weights – they were needed for understanding reactions and calculating ingredient ratios. In the early 1800s, English chemist John Dalton took hydrogen as a unit and used it to compare other elements' relative weights. Swedish pioneer Jacob Berzelius did something similar but using oxygen as a standard, in 1818.

The most impressive of such efforts was the work of Edward Morley, an American chemist who, in 1895, announced an incredibly precise value for the atomic weight of oxygen compared to hydrogen. He had calculated the figure through an elaborate series of experiments over the course of 11 years. Since so many other elements' weights were found by synthesizing oxides – an element plus oxygen – it was a great step forward.

Early periodic tables tried to order the elements using atomic weights. Chemists today still measure atomic weight, also called relative atomic mass, in relation to the mass of an atom of carbon-12. The calculation takes into account the average mass of all an element's isotopes and allows scientists to measure reactants and products accurately.

▲ *British-American astrophysicist Cecilia Payne discovered the composition of stars in 1925.*

THE SOURCE OF OUR ELEMENTS

Where do all the elements on Earth come from? To answer the question takes a real star. In 1925, a British-American scientist called Cecilia Payne (later Payne-Gaposchkin) first realized that the Sun was mostly made of hydrogen and helium. She made the finding during her Ph.D research – and came into immediate conflict with the prevailing view, which was that stars had a similar chemical make-up to Earth. But Payne was right. When doubters caught up with her, four years later, the abundance of hydrogen fuelled many more discoveries.

▲ *Payne deduced that hydrogen was by far the main element in the stars, and was therefore the most abundant element of all.*

English astronomer and physicist Arthur Eddington had already suggested that, within stars, lighter elements might be squeezed together into heavier ones. He published the work in 1926, but nobody fully understood how it might happen. German-born physicist Hans Bethe then worked out how stars produce energy. The process within our Sun, for example, involves four hydrogen nuclei fusing to form one helium nucleus. The mass of helium is less than the sum of the hydrogen nuclei, and the difference in mass turns into energy.

Hydrogen and helium had been around since the beginning of the universe. For the first few minutes after the Big Bang, it was pretty much all there was. As things cooled and stars formed and ignited, element production could begin.

▲ Fred Hoyle.

▲▶ Iron (above) and nickel (right) are the heaviest elements our Sun can make.

STELLAR INGREDIENTS

But how would stars go about making elements heavier than helium? English astronomer Fred Hoyle published the first research relating to making more elements within stars in 1946. His 1954 paper entitled *Synthesis of the elements between carbon and nickel* then explained how the stellar ingredients hydrogen and helium could turn into the most common elements on Earth. With colleagues Margaret Burbidge, Geoffrey Burbidge and William Fowler, Hoyle expanded the idea to account for even heavier elements.

Once a star's core turns to iron and nickel, it can no longer generate energy through fusion. It may then turn supernova, exploding with terrific energy – and making many more, heavier

▲ *Margaret Burbidge, Geoffrey Burbidge and William Fowler worked with Fred Hoyle on a landmark 1957 scientific paper about how the stars synthesize elements. It is known as B^2FH after its authors.*

elements. As atoms are blown outward, they are bombarded with neutrons that they capture in their nuclei, turning into new protons. As a result, iron can turn into gold; gold into lead and so on, up to the heaviest star-formed element, uranium.

And so, the elements blast out into space. Some of the matter regroups and is recycled into new stars. Some arrives here on Earth – the way it has done for billions of years. Perhaps it seems like a bit of stardust to the supernova. To us, these elements are the stuff of life.

ELEMENTS MADE IN A SUPERGIANT STAR

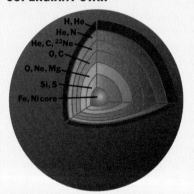

H, He
He, N
He, C, ^{22}Ne
O, C
O, Ne, Mg
Si, S
Fe, Ni core

Super red giants fuse hydrogen to helium, helium to carbon and so on up to iron and nickel, before turning supernova and making even heavier elements.

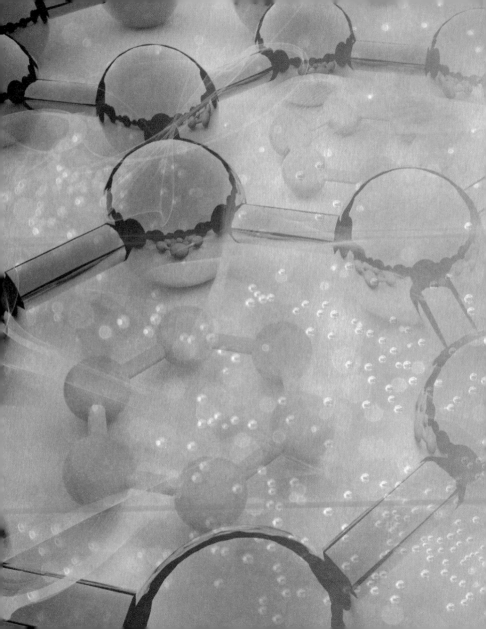

CHAPTER 1

DEVELOPING THE PERIODIC TABLE

HOW DID FIZZY WATER GIVE THE
DEVELOPMENT OF CHEMISTRY A LIFT?

WHAT DID LAVOISIER LEARN BY BURNING A DIAMOND?

WHY WAS MENDELEEV'S PERIODIC TABLE
JUST NOT QUITE RIGHT?

MAKING SENSE

Patterns are how we make sense of the world around us. When we are babies, the vague realization that our gurglings or wriggles lead to more-or-less predictable outcomes is one of the ways we learn. Our earliest ancestors noticed sequences of day and night, warm seasons and cold, the constellations shifting slowly across the skies, and built an understanding of nature.

Today, even though the world often seems incomprehensibly complex, we all use patterns in a scientific way. We sense a system in the way things work, use it to predict an outcome, and adjust our expectations for next time according to the results. Our understanding is that, despite complication and chaos, there is order and meaning – and we are determined to find it.

The Greek thinker Empedocles suggested 2,500 years ago that earth, fire, air and water were the indestructible elements that lay within all substances. The idea was that every existing material had properties of the four elements, and could change from one substance to another if the proportions of the elements were altered. Empedocles' notions were adopted and popularized by Aristotle, who was born a hundred years later.

▶ *Fire, water, earth and air were the four elemental ingredients of all matter, according to Greek thinker Empedocles.*

▶ *Jabir ibn Hayyan, the 8th-century chemist, had influential ideas about classifying substances, and carried out wide-ranging practical experiments.*

The four-element system was compelling enough to survive for centuries. In the year 790 CE, in Iran, Jabir ibn Hayyan additionally suggested that metals were made from two elements, sulfur and mercury. Known as Geber in the West, his ideas reached Europe via translations into Latin in the 11th century, as did many other Arabic texts. The charismatic Swiss doctor Paracelsus added salt to the duo of sulfur and mercury in the 1500s and called them the three principles, from which nature made all things.

▲ *Paracelsus' writings are full of mystical cures and alchemy, but he was one of the first people to bring chemistry into medicine.*

Elements then meant something very different to today. Superstition often flourished as alchemists used the different elements to try to harness forces to turn base metals into gold or give eternal life. But people were looking for patterns by which to make some sort of sense of things – an impulse that led ultimately to great sweeps of human development. In terms of the chemistry and physics of our world, a gradual recognition of patterns and systems is how we have come to understand everything from the solar system to our own bodies.

REMAINING QUESTIONS

That is not to say that everything is now sorted out, tidied up, cut and dried. The search for answers continues in science, engineering and medicine. What is the process by which a healthy brain builds itself? How can we make sense of black holes? Why does the placebo effect work? When will the whole world have clean water? What is consciousness? Will fusion power mean unlimited energy? Why do cells in the body sometimes turn cancerous? When will we be able to teleport ourselves from one place to another? How would we recognize signals from aliens?

Progress in areas like these is vital for the wellbeing of our populations and planet. But to be able to tackle questions of such complexity is in itself a privilege. Our research today is based on thousands of discoveries and breakthroughs from the past, fundamental principles, laws, relationships and observations. Every scientist working now is able to draw on a hard-won and continuing history of proposals and proofs, missteps and corrections, ideas that withered away – and ideas that have come to fruition.

▲ *Venkatraman Ramakrishnan.*

▲ *Youyou Tu.*

PRIZE-WINNING PATTERNS

New discoveries are helping to tackle significant questions, for example:

▲ Ada Yonath.

- How do our cells build the proteins we are made of? In 2009, the Nobel Prize for Chemistry was awarded to Venkatraman Ramakrishnan, Thomas Steitz and Ada Yonath for working out the detailed structure and mechanism of the ribosome, a complex machine that reads protein recipes and then creates them.

- What are the new materials that could transform electronics and battery technology? An entirely new form of carbon, called graphene, was the subject of the 2010 Nobel Prize for Physics. Awarded to Andre Geim and Konstantin Novoselov, the prize celebrated a material that is at the same time incredibly thin, strong, conductive and transparent.

- Where are the medicines that can treat the world's biggest killer diseases? Chinese chemist Youyou Tu won the 2015 Nobel Prize for Medicine for her work to extract a substance called artemisinin from sweet wormwood, a plant used in Traditional Chinese Medicine. Artemisinin is being used to inhibit the malaria parasite and has improved the lives of millions.

▲ Thomas Steitz.

PATTERNS OF MATTER

The periodic table is where patterns of matter in our world are clearly laid out. It brings together all the known pure substances in a systematic way that helps summarize their intrinsic nature and visualize their properties.

The first two columns contain elements that become increasingly reactive as you descend the table. They are the very lively **ALKALI METALS** in the first column, and the slightly calmer **ALKALINE EARTH METALS** in the second.

In the centre of the table are the **TRANSITION METALS**, which form a broad block of mostly tough and useful stuff with some precious or unusual metals, such as mercury, in the mix.

Beneath these – actually belonging between the alkaline earths and the transition metals, but placed below for ease of presentation – are two groups that together form the rare earth metals. The upper row contains the **LANTHANIDES**, dense shiny metals that slowly tarnish in the air. The lower row contains the **ACTINIDES**, which are all radioactive.

The next four vertical groups, headed by **BORON**, **CARBON**, **NITROGEN** and **OXYGEN**, tend to be metallic in nature at the lower left-hand corner, and non-metallic toward the upper right-hand corner. Oddly enough, although there are so many metals in the periodic table, it is among the non-metals that we find the most abundant elements in our world and universe.

A double diagonal strip runs through this group-of-groups, consisting of the

1 H

3 Li	4 Be

11 Na	12 Mg

19 K	20 Ca	21 Sc	22 Ti	23 V
37 Rb	38 Sr	39 Y	40 Zr	41 Nb
55 Cs	56 Ba		72 Hf	73 Ta
87 Fr	88 Ra		104 Rf	105 Db

57 La	58 Ce	59 Pr
89 Ac	90 Th	91 Pa

Hydrogen

Alkali Metals

Alkaline Earth Metals

Transition Metals

Lanthanides

Actinides

The Boron Group

The Carbon Group

The Nitrogen Group

The Oxygen Group

The Halogens

The Noble Gases

						5 B	6 C	7 N	8 O	9 F	10 Ne
						13 Al	14 Si	15 P	16 S	17 Cl	18 Ar

2 He

25 Mn	26 Fe	27 Co	28 Ni	29 Cu	30 Zn	31 Ga	32 Ge	33 As	34 Se	35 Br	36 Kr
43 Tc	44 Ru	45 Rh	46 Pd	47 Ag	48 Cd	49 In	50 Sn	51 Sb	52 Te	53 I	54 Xe
75 Re	76 Os	77 Ir	78 Pt	79 Au	80 Hg	81 Tl	82 Pb	83 Bi	84 Po	85 At	86 Rn
107 Bh	108 Hs	109 Mt	110 Ds	111 Rg	112 Cn	113 Nh	114 Fl	115 Mc	116 Lv	117 Ts	118 Og

| 61 Pm | 62 Sm | 63 Eu | 64 Gd | 65 Tb | 66 Dy | 67 Ho | 68 Er | 69 Tm | 70 Yb | 71 Lu |
| 93 Np | 94 Pu | 95 Am | 96 Cm | 97 Bk | 98 Cf | 99 Es | 100 Fm | 101 Md | 102 No | 103 Lr |

semi-metals that are so useful in semi-conductors, the basis of our computers.

In the penultimate column of the periodic table, the **HALOGENS** are highly reactive at the top of the group and decreasingly so as you go down. The final column contains the **NOBLE GASES**, which are all colourless gases and uniformly nonreactive.

So, how did people put together the pieces to build the periodic table? What were the patterns they perceived that helped create the table in its current form? And how did their insights help further our understanding of how the world works?

THE POWER OF OBSERVATION

Jan Baptist van Helmont lived in a volatile period for chemistry. Born in 1580 in Brussels, he was well aware of the alchemical teachings of Paracelsus. On the other hand, he wanted to learn by observation and experimentation, as did contemporaries including Galileo Galilei in Italy.

Van Helmont tackled a common misconception of the time: that plants grew by eating soil. He grew a willow plant in a pot by itself, providing it only with water. After five years, the plant had gained hugely in weight while the soil had lost only a few grams.

For his next trick, van Helmont also demonstrated that if you dissolve silver in acid, you can recover all the precious metal again by using iron. It seemed like the beginning of the end for alchemy – and the start of new ways to classify substances.

THE DECLINE OF ALCHEMY

Robert Boyle would help hasten alchemy's demise. In 1641, when he was 14, the young Anglo-Irish aristocrat travelled to Italy. Galileo was living in Florence under house arrest because of his ideas about the Sun, rather than the Earth, being at the centre of the solar system.

Boyle read and appreciated Galileo's mathematics, which shed light on the fundamental motion of planets and pendulums alike. He returned home and a few years later moved to Oxford and

▲ Italian physicist and astronomer Galileo Galilei published this radical diagram in 1632, showing the six known planets orbiting the Sun.

set up a laboratory to study vacuums and gases. While working with Robert Hooke, Boyle discovered a mathematical relationship between the volume and pressure of a gas – when multiplied together they are constant: when one increases, the other decreases. He also found that sound cannot travel through a vacuum – and fire cannot burn in one.

SCEPTICAL CHEMISTRY

In 1661, Boyle published his most influential book, *The Sceptical Chymist*, applying his rational eye to the study of chemistry. He correctly identified that elements were substances that could not be decomposed further. He emphasized the need to repeat experiments and gather plenty of data before drawing conclusions – and he rejected both Aristotle's and Paracelsus' elements.

Boyle moved to live at his older sister Katherine's home in London's fashionable Pall Mall in 1668. Katherine was interested in science in her own right, and commissioned work to build a new laboratory, which Robert used for the next 30 years.

▼ Alchemy played an important part in Robert Boyle's thinking, but he also believed in practical chemistry and the recording of results.

▲ Robert Boyle experimented on air, successfully showing that sound from a bell did not travel through a vacuum.

27

SPA WATERS AND SCIENCE

Bubbling mineral waters and springs have sustained the search for patterns in chemistry over the centuries.

Italian doctors studying health-giving springs in the 13th century were continuing a line of thought from antiquity. Ancient Greek writers spoke of evaporating mineral waters to examine what remained. Giacomo de Dondi recommended in the 14th century that you should test such a residue on hot coals to release any hidden odour. Bartolomeo Montagnana, who died in 1460, analyzed spa waters by distilling out their mineral contents, a process by which different contents of the water could be identified by their boiling points. His contemporary, Michael Savonarola, developed ways to identify sulfur in mineral water residue because it burned with a particular colour flame.

Jan Baptist van Helmont experimented with mineral water in the early 1600s, discovering a number of gases including carbon dioxide, which he called *spiritus sylvestris*. Robert Boyle published a study of London well water in 1685, setting out for the first time a clear method for analyzing its chemical composition. Friedrich Hoffmann, a German doctor, wrote about "steel waters" to strengthen limbs and "bitter purging waters" to help with fever, basing his chemical tests on those of Boyle.

Peter Shaw, one of King George II's doctors, translated Hoffmann into English in 1731 and went on to develop a method for analyzing water based on his understanding that mineral waters contained salts, earths, sulfurs and fumes – this was the result of dissolved minerals from nearby rocks.

Joseph Priestley was a polymath from northern England, who studied mineral water hoping, like many, to recreate

► What's in your drinking water? The study of springs and spas helped advance chemistry.

► *The thermal baths at Vichy in France were popular in Roman times, and had their heyday in the 1930s.*

▼ *Torbern Bergman's book De Analysi Aquarium gave the first comprehensive analysis of mineral water.*

healing properties. He studied a variety of gases, deciding that there were three types of air: fixed, alkaline and acid. In 1774 he isolated an entirely new kind of air, which turned out to be oxygen. He even succeeded in producing the first artificial fizzy drink in 1772 – water impregnated with fixed air, as he called carbon dioxide. It was a sensational success.

When Swedish chemist Torbern Bergman wrote a tome on mineral water analysis in the 1770s, he called it "one of the most difficult problems in chemistry". Nonetheless, the study of spring water had yielded new elements, new techniques – and even a new industry, in soft drinks.

EARLY CLASSIFICATION

One way to organize substances was by their tendency to combine – a property known as their chemical affinity.

A French chemist and doctor called Étienne François Geoffroy produced a table of affinities in 1718, which used symbols to represent acids, alkalis, metals and water. At the top of each column was a substance with which all the items below could combine. For example, the substance at the top of the third column was nitric acid, which the table recorded would combine with the metals below: iron, copper, lead, mercury and silver.

What is remarkable is that the order in which these metals appeared in the 300-year-old table is similar to a reactivity series today.

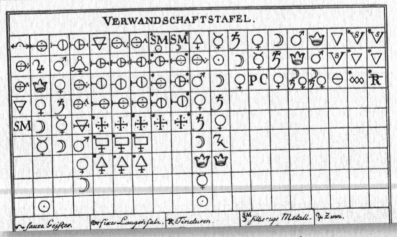

▲ When he created his table showing the ways chemicals interacted, Étienne Geoffroy used symbols from alchemy to represent substances such as sulfur, iron and gold. But there was some good science in his observations of patterns in reactivity.

ALCHEMY IN THE ARABIC-SPEAKING WORLD

Investigation and discovery flourished in the Arabic-speaking world from the 7th century. Despite the continuing mystical aspects of alchemy, there was much practical chemistry we might recognize today. Chemists worked in laboratories using crucibles, scissors and spoons; furnaces, funnels and filters. Most interestingly, they had perfected the process of distillation – the process of separating a mixture into its component parts by using their different boiling points. Glass alembics from as early as the 8th century exist in museum collections today, their name, al-inbiq, a reminder of their origins.

Arabic-speaking alchemists defined elementary properties of substances in terms of heat, cold, dryness and moisture. Their ultimate goal was to gain wisdom rather than riches, so the

▲ *Chemists mix remedies in this Arabic translation of the* Materia Medica *by 1st-century scholar Dioscorides.*

◀ *Rosewater was one product distilled during the 8th century.*

rules of the alchemist were deliberately cloaked in mystery to avoid them being used carelessly. A famous text by doctor and alchemist Al-Razi was called *Secret of Secrets*.

But their records include the first descriptions of substances including sulfuric and nitric acid, along with soda (which they called al-natrun) and potassium (al-qali). Other words with an Arabic legacy include alcohol (al-kohl) and elixir (al-iksir). The researchers of this period may not have found all the secrets of life, but their legacy certainly survives.

Such a useful table gained huge popularity among researchers. A dozen other chemists published affinity tables in the later 18th century, the largest of all by Torbern Bergman, the Swedish professor of chemistry who also studied mineral waters. In 1775 he developed a 50-column, 50-row chemical affinity table showing thousands of chemical reactions. Up until this time, chemists had always assigned alchemical symbols to particular substances, but Bergman used letters to denote the single or combined species in the table: A, B, C as individual substances and AB, AC as compounds.

A LIST OF ELEMENTS

It was a pioneering French chemist, Antoine Lavoisier, who would make the first giant steps toward a true understanding of elements. He was fastidious about accurate measurement, and one of his theories was that mass was conserved in reactions – whether or not you could still see the products. In 1772, he performed an incredible experiment using a huge vinegar-filled lens, in which he burned a diamond apparently to nothing, then weighed the jar and found it unchanged. The same occurred when he burned charcoal, and he deduced that

◄ By using a huge lens that focussed the rays of the Sun, Antoine Lavoisier burned a diamond until it seemed to disappear – but its mass was conserved.

▲ *French scientist Antoine Lavoisier worked on experiments that put modern chemistry on the map.*

He had already confirmed, in 1777, that sulfur was an element – a substance that could not be broken down any further. But it was a huge surprise to many people to learn that Lavoisier had shown water was *not* an element – he could make it by combining oxygen with hydrogen.

In 1789, Lavoisier published his groundbreaking book, *Elementary Treatise on Chemistry*. In it, he listed substances he said were chemical elements, including oxygen, nitrogen, hydrogen, sulfur, phosphorus, carbon, antimony, arsenic, cobalt, copper, gold, iron, manganese, molybdenum, nickel, platinum, silver, tin, tungsten, bismuth, lead, manganese and zinc.

the substances were forms of the same element. He named it carbon.

Lavoisier investigated the science of burning, which most people believed occurred because substances contained a component of fire called phlogiston. Lavoisier learned of work on gases by Joseph Priestley, and also an experiment by reclusive English scientist Henry Cavendish in 1766, in which he had created a gas he called inflammable air. Lavoisier realized it was oxygen rather than the fictitious phlogiston that made things combust, and, in 1779, he named the gas oxygen.

▲ *Lavoisier made the first extensive list of elements, including cobalt, copper, gold, tin and zinc. He confirmed carbon as the name of an element with two forms: charcoal (above) and diamond.*

These were all correct. He did, however, also include a substance called caloric, which he thought explained why things heated up – a fiction that lasted until James Joule, in 1849, showed how energy could be converted from one form to another. However, Lavoisier's work had elevated chemistry to an entirely new level.

SIMPLICITY FROM ATOMIC THEORY

Elements could be invisible gases, they could be metallic solids...so what were these substances like at a fundamental level? John Dalton was a self-taught scientist and lecturer in Manchester, northern England. In 1802, he published some new ideas relating to gases, including useful principles and laws that brought him to wider notice. He independently made discoveries about the pressure of two gases mixed together that was made famous by the French chemist Joseph Louis Gay-Lussac, in 1808.

Dalton agreed with many scientists of the day that matter was made from particles too small to see. But his breakthrough was to realize that all atoms of a particular substance are identical, and can be identified by their weights. In a book of 1808 he revealed his atomic theory. It was like a new language for chemistry.

▲ John Dalton's scientific interests included atomic theory, meteorology and also colour blindness.

▼ Dalton devised a system of symbols for 20 elements and worked out atomic weights for them.

ELEMENTS

Symbol	Element	wt.	Symbol	Element	wt.
⊙	Hydrogen.	1	⊕	Strontian	46
⊖	Azote	5	⊗	Barytes	68
●	Carbon	5½	Ⓘ	Iron	50
○	Oxygen	7	Ⓩ	Zinc	56
⊘	Phosphorus	9	Ⓒ	Copper	56

Dalton worked out the atomic weights of 20 substances, to which he assigned symbols, and also stated that these elements would combine together in simple, whole-number ratios. In a pamphlet promoting his book, Dalton wrote that it "tends to place the whole science of chemistry upon a new, and more simple, basis than it has been upon heretofore."

ATOMS TO MOLECULES

Around the same time as Dalton was working in England, an Italian scientist called Amedeo Avogadro was also wondering about atoms. In his twenties he had already produced his first scientific paper, which used the newly invented voltaic pile, or battery, to investigate salt solutions.

He studied John Dalton's work on atoms, and also Joseph Louis Gay-Lussac's ideas about how gases react. Then, in 1811, he published a paper containing the idea that we now know as Avogadro's law: that equal volumes of all gases at the same temperature and pressure will contain the same number of atoms. Avogadro was the first to realize that elements could exist as molecules rather than as individual atoms – as hydrogen, oxygen and many other elements indeed do.

▲ Amedeo Avogadro understood that elements could form molecules rather than existing only as individual atoms.

⊕	Sulphur	13	Ⓛ Lead	90
◐	Magnesia	20	Ⓢ Silver	190
⊖	Lime	24	ⓓ Gold	190
◉	Soda	28	Ⓟ Platina	190
⦀	Potash	42	⚙ Mercury	167

MORE NEW ELEMENTS

In Sweden, scientists were successfully seeking new elements. Jacob Berzelius was a trained doctor whose first love was chemistry. Like Avogadro in Italy and Humphry Davy in England, Berzelius wanted to apply electricity to chemicals using battery technology. He learned how to split salts, and contributed to elemental discoveries by Davy via their correspondence.

▲ Chlorine, bromine and iodine have similar properties.

In 1803, Berzelius discovered a new element, cerium, and also found thorium in 1815, selenium in 1817 and later isolated silicon. But his masterwork was a huge table of atomic weights, begun after encountering John Dalton's work and published in 1818. He could not quantify atoms directly, so the idea he pursued was to use the element oxygen as a standard against which to measure other substances. Beginning with gases, then moving on to metals, Berzelius purified and analyzed hundreds of compounds and found remarkably accurate weights for many elements. He also rationalized the chemical symbols used for the elements – calling each by a one- or two-letter short form based on their Latin name.

▲ A Swedish postage stamp of 1979 remembered pioneering chemist Jacob Berzelius on the 200th anniversary of his birth.

ELEMENTS IN THREES

Lists of elements were getting longer. A breakthrough in organizing them into manageable subsets appeared in 1817. A German chemist called Johann Wolfgang Döbereiner showed that the weight of the element strontium was half-way between those of calcium and barium. He later managed to make a similar link between other chemical triplets – chlorine, bromine and iodine; and lithium, sodium and potassium – which he noted had similar properties. Döbereiner published a report in 1829, referencing the efforts of Berzelius in working out the atomic weights of bromine and iodine.

Döbereiner made his calculations based on atomic weights, prior to Avogadro's insight into the fact that some elements existed as molecules – but, even so, his averages did relate substances together in the right ways.

ELEMENTS IN EIGHTS

A sweet-sounding answer to the question of elemental patterns was English scientist John Newlands' idea in 1864. He noticed that by arranging the elements into order of their atomic weight, their properties repeated every eighth element. It was like the patterns of notes on a piano and he wrote that the elements "stand to each other in the same relation as the extremities of one or more octaves of music."

He presented his work to the Chemical Society of London, but they did not take to the idea, to Newlands' great disappointment – one problem was that, further up the scale, the properties of the elements did not really continue the pattern closely. It turned out this was because elements were missing, but Newlands did not want gaps in his table and chose to close them up.

◀ Johann Wolfgang Döbereiner spotted links between triads of substances, a step forward in classifying the elements.

ELEMENTS IN SIX FAMILIES

German scientist Julius Lothar Meyer also noticed that if elements were placed in atomic weight order, their chemical and physical properties repeated regularly. He produced an early version of his periodic table in 1864. It contained 28 elements, grouped into six families that reflected their valence, or power to combine with other substances – a far-sighted move. By 1868 he had improved the table by adding the transition metals, but it was not published until 1870.

◀ *Julius Lothar Meyer.*

 ▲

Mendeleev predicted the existence of elements with properties that matched those of gallium (left; element 31, discovered in 1875) and germanium (above; element 32, discovered in 1886).

THE PREDICTIVE PERIODIC TABLE

In 1869, Dmitri Mendeleev produced his periodic table, which contained 65 elements arranged in the order of their atomic weights. He had also spotted a repeat in chemical properties after every eight elements, and arranged them accordingly in his table, fitting them into 12 rows, or periods. Mendeleev did not close up any gaps in which there seemed to be missing elements, and indeed even predicted the properties of the substances that should fill the holes.

When other scientists discovered some of the missing elements with the characteristics Mendeleev had foreseen, it was excellent confirmation of his work. Mendeleev had also had the confidence to say that some of the atomic weights of known elements must have been measured incorrectly, and again, he was proven right.

▲ *Dmitri Mendeleev.*

TWENTIETH-CENTURY CHEMISTRY

It had been a brain-bending, sometimes maddening, puzzle for chemists to uncover the pieces of the periodic table and then fit them all together, and Mendeleev's table had proven both insightful and useful. Yet in the early 1900s, it was clear that the jigsaw was still not right. Chemists could see that the position in the table that the element's atomic weight indicated did not always make the properties fit correctly.

The work of a young chemist called Henry Moseley would provide the evidence to explain what was going on.

PERIODIC PIONEERS
DMITRI MENDELEEV

Brilliant, bushy-haired, bad-tempered. The chemist Dmitri Mendeleev seems like an archetypal eccentric scientist. But then…it is a caricature that seems so closely based on Mendeleev, it is practically a tribute.

Born in the Russian province of Siberia in 1843, Mendeleev was the youngest child in a large family. His father had studied in the city of St Petersburg and worked as a teacher. However, he died when Dmitri was 13. Two years later, the glass factory that then provided the family's income burned down.

Mendeleev's resourceful mother could see education was Dmitri's only route out of poverty. The two of them rode on horseback 2,000km (1,200 miles) to Moscow, but Dmitri was rejected by the university because he was not a local student. They had to continue a further 650km (400 miles) to reach the capital city St Petersburg, where they managed to secure Dmitri a place in his father's former college.

Study suited Mendeleev. He progressed rapidly, and at 20 he was already publishing research papers. Despite his sometimes brusque approach to colleagues – and a bout of tuberculosis – he still graduated at the top of his year. He gained a master's degree in chemistry in 1856,

▲ *The USSR produced this stamp in 1971 to honour chemist Dmitri Mendeleev.*

then won a grant to pursue research in Heidelberg.

This was where some of the most exciting chemistry was developing. Robert Bunsen was discovering new elements with spectroscopy, and Mendeleev learned about the technique from him. In 1860, at the first-ever international chemistry conference, in nearby Karlsruhe, Mendeleev heard the discussions focus on one issue alone: how to standardize chemistry.

CLASSIFICATION

He returned as a lecturer to St Petersburg, determined to see chemistry flourish – particularly in his home country. At great speed, he wrote first one textbook, then a second – which is when he had his revelation. He wanted to explain to students about the organization of the elements, so he wrote the names and profiles of the 65 substances then known onto cards. His feverish attempts to find a satisfactory way to classify them left him exhausted, and he fell asleep.

In the story he would tell, he said that in a dream he saw a table in which all the elements fell into place as required. He wrote it all down when he woke up. His textbook was published two weeks later.

▲ Manuscript of Mendeleev's first periodic system of elements, 17 February 1869.

PARTICLE PLACEMENT

In 1913, working at Oxford University, Henry Moseley was intrigued by a new way of considering the placement of elements in the periodic table. The idea was suggested by an amateur physicist from the Netherlands, Antonius van den Broek, who had read research about the atomic nucleus by Moseley's former boss, Ernest Rutherford. Van den Broek thought that rather than placing elements by atomic weight, it should be by atomic number. While atomic numbers had only been considered as placeholders before, van den Broek linked them to the amount of charge inside the nucleus of each atom.

Moseley knew that X-rays might be able to help prove this idea one way or the other. Work by William Lawrence Bragg at Cambridge University, and his father William Henry Bragg, had recently shown that X-rays could reveal the positions of atoms within a crystal through interpreting the spectrum of diffraction lines they produced. Moseley set up an experiment to shoot high-powered electrons at the elements and measure the X-rays that they emitted – a novel use of X-ray spectroscopy. He was amazed to find a systematic relationship between the frequencies of X-rays emitted and the element's atomic number, just as van den Broek had suggested.

▲ William Lawrence Bragg.

◄ *Iodine.*

▲ *Tellurium.*

The discovery made most sense, Moseley realized, if the atomic number corresponded to the number of protons in the nucleus. For the first time, Moseley had evidence that elements were defined by their proton number – the way we do it today.

ATOMIC NUMBERS

Suddenly Moseley could supply the logic for the choices earlier researchers had made in order to make the periodic table reflect the patterns in reality. Two transition metals, cobalt and nickel, had very similar atomic masses. The hard-headed Mendeleev had fitted them into the table by their physical and chemical characteristics rather than strictly by mass. Iodine and tellurium, too, had always been located out of order so as to group the two with other similar

elements: iodine among the halogens and tellurium alongside another semi-metal, selenium. But by using atomic number to order the elements, everything made sense. Mendeleev had been right about his groupings – correct about sticking with patterns in the periodic table.

Moseley was also able to see that there were gaps in the periodic table. Elements were missing with atomic numbers 43, 61, 72 and 75. All these elements were found in the years that came after: technetium (which Mendeleev had also foreseen), promethium, hafnium and rhenium.

The wisdom of the search for patterns had never been proven more clearly.

THE BRILLIANT HENRY MOSELEY

The few, well-known photographs of Henry Moseley show him as a neat-haired young man, smartly dressed, with a hint of a moustache so familiar from the First World War period.

What is so sad is that he, like many men of his generation, died in that conflict in 1915. He was only 27, and lost his life to a sniper's bullet at Gallipoli as he was sending an order. Nobel Prizes in physics and chemistry went unawarded in 1916. Many have speculated that Moseley would have been a prize-winner if he had

▲ *Henry Moseley in his Oxford University laboratory.*

lived, and Isaac Asimov wrote later that, "In view of what he [Moseley] might still have accomplished…his death might well have been the most costly single death of the War to mankind generally."

Henry Moseley did not have to enlist to fight; he joined up as a volunteer and went against his family's wishes that he continue his scientific research.

In a short career, Moseley had made enormous contributions already. He had graduated from Oxford University and then gone to join the Nobel Prize-winning scientist Ernest Rutherford at Manchester University to work on radioactivity. One of the spin-offs of the work he did on radium was the atomic battery that is now used in long-life applications like spacecraft.

As well as his breakthrough work on atomic numbers back at Oxford, Moseley's X-ray work yielded a new, non-destructive process for analyzing unknown samples. Essentially, you bombarded the substance with high-powered electrons and looked at the resulting X-ray frequencies, which corresponded specifically and reliably with the elements inside. X-ray spectroscopy is now a standard technique.

Physicist Robert Millikan said in 1923 that Moseley "threw open the windows through which we can now glimpse the subatomic world with a definiteness and certainty never even dreamed of before."

1
H
Hydrogen

CHAPTER 2

HYDROGEN

WHY IS THE NAME "WATER CREATOR"
SO PERFECT FOR HYDROGEN?

HOW DID HYDROGEN REVEAL
THE HIDDEN STRUCTURE OF THE ATOM?

COULD FUEL CELLS GIVE US A GREENER WAY TO FLY?

HOW HYDROGEN UNLOCKED ATOMIC STRUCTURE

It is incredible to think that all the matter surrounding you – the chair you are sitting on, the drink in your hand, the air you are breathing – is made of atoms that consist of just three kinds of particles. Every single physical substance is a combination of protons, neutrons and electrons, the number and arrangement of which makes the difference between being a bluish-grey metal called silicon (14 protons and 14 electrons), or a toxic non-metal called phosphorus (with 15 of each).

As the lightest and simplest element, hydrogen has played a central role in experiments and theories relating to atomic structure. In 1913, Danish physicist Niels Bohr suggested a vivid model for hydrogen, which brought together all the available evidence at that time. He said that inside each atom of hydrogen, a single negatively charged electron orbited a positive nucleus, a bit like a planet going around the sun. And he added that the electron could inhabit certain energy levels, or shells, and they could jump to a higher level as they absorbed energy in the right amount, and drop back to their ground state by emitting energy.

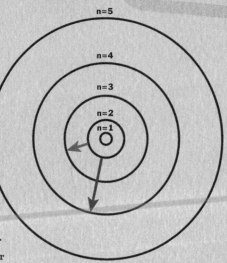

▲ In Bohr's planetary model of the atom, electrons could jump between levels as they absorbed or emitted energy.

Fast Facts
HYDROGEN [H]
ATOMIC NUMBER 1

Character: Hydrogen is a colourless and highly flammable gas. It accounts for 88 per cent of all atoms, and 62 per cent of the atoms in your own body. Under immense pressure hydrogen may become a liquid metal – a substance scientists believe exists in the interior of Jupiter.

Discovery: Alchemists of the 1500s knew that adding iron filings to acid resulted in bubbles of gas. But the first person to recognize that this was a new substance was British chemist Henry Cavendish in 1766, who named it inflammable air.

Name: Hydrogen means *water creator* in Greek. In 1783, French chemist Antoine Lavoisier confirmed earlier findings that burning this gas in air produced water.

World sources: Ninety per cent of hydrogen is produced from fossil fuels by passing superheated steam over hot coke, a form of carbon, or methane gas. A small proportion comes from splitting water using electrical energy, a technique called electrolysis, which can be a zero-emission process if done using renewable energy.

▲ *Planet Jupiter.*

British chemist Henry Cavendish. ▶

▲ *Hydrogen power is considered a "clean" source of energy and is being used to fuel buses and cars.*

INSIDE THE ATOM

Niels Bohr built his planetary model on evidence found around the turn of the 20th century for the inner contents of atoms. While the word atom originally came from the Greek for "indivisible", experiments had shown that there was a great deal going on inside them.

First of all came electrons. A British physicist called J J Thomson was interested in electricity and magnetism, including the way in which electricity passes through a gas. In 1897, he was working with a Crookes tube, a glass vessel with only a trace of air left inside, to which he applied a voltage to make a ray travel through the tube. He managed to show that he could deflect the ray by using an electric or a magnetic field. This finding indicated that the ray was made from physical particles carrying a charge, rather than being an electromagnetic beam, as some scientists proposed.

But what might these particles be like? The direction of deflection told Thomson that they must carry a negative charge. By measuring how much they had deflected, he could also measure the ratio of charge to mass which the particles must have.

◀ *JJ Thomson at the age of 53.*

CROOKES TUBE

A Crookes tube is a glass vessel with nearly all the air pumped out of it, fitted with two electrodes. Invented in the 1870s, this clever device helped illuminate the physics of electrons.

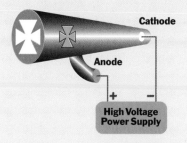

When a high voltage is applied between the electrodes, free electrons in the tube start to accelerate toward the positive electrode, knocking more electrons out of the air particles as they go. The air particles, now with a positive charge, are attracted toward the negative electrode called the cathode, and collide into it. This releases a stream of electrons from the cathode's metallic surface called a cathode ray.

William Crookes, after whom the tubes are named, worked out how to evacuate the tubes to such low pressures that the cathode ray electrons could travel the length of the tube without bumping into many air particles. They hit the far end of the tube with enough energy to make the glass glow. An object blocking the ray, such as a Maltese cross shape, would leave a shadow.

▼ *An original vacuum tube used by JJ Thomson to discover the electron in 1897.*

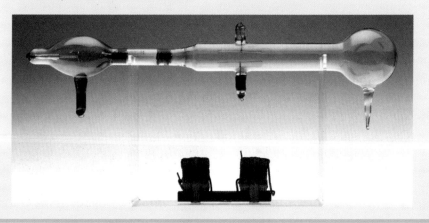

ELECTRONS AND THE PLUM PUDDING MODEL

J J Thomson proposed that these new particles were tiny, negatively charged constituents of every sort of atom. He called the little lightweight particles corpuscles, but others preferred the name previously suggested by a Scottish scientist for a unit of electrical charge – electrons.

▲ Thomson's model of the atom suggested the electrons were distributed like the plums in a pudding.

This was the first breakthrough that indicated that atoms contained smaller elementary particles, and were not indivisible units of matter after all. But because atoms are neutrally charged, the finding of negatively charged electrons required more thinking about a balancing positive charge. Thomson suggested that the tiny electrons might be held within some sort of positively charged paste that also accounted for the mass of the atom. His model – honouring the British love for stodgy desserts – was of a plum pudding, with the electrons as the fruit studding the positively charged sponge.

The plum pudding model worked well until 1911. Then, Thomson's former student Ernest Rutherford served up some unexpected experimental results. He had fired alpha particles – these were

▲ Japanese physicist Hantaro Nagaoka suggested a planetary model for the atom, based on Saturn's rings, before Niels Bohr's proposal.

Ernest Marsden (left) worked with Ernest Rutherford (right) on the gold foil experiment that revealed the atomic nucleus. ▶

positively charged particles spontaneously emitted by a radioactive source – at a sheet of gold foil, and looked at how the particles were affected. The plum pudding model predicted that the particles would be deflected a little bit as they passed through the positive substance of the gold atoms.

THE NUCLEUS

But what Rutherford saw was very different and surprising. Most of the particles were completely unaffected as if they had passed through empty space – while some actually bounced off the foil. The only explanation was that these particles had hit something very heavy indeed.

What Rutherford had shown was that the vast majority of the mass was concentrated into a tiny volume at the centre of the atom – the nucleus. He realized that he must be dealing with a new elementary particle. Protons, as he later named them, were positively charged particles that made up the nucleus, which he imagined the electrons surrounded in a kind of cloud.

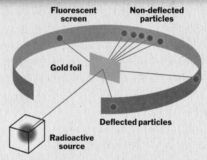

▲ In the gold foil experiment, a few alpha particles were deflected by large angles because they had hit the gold nuclei.

Niels Bohr took up the story with his planetary model of the hydrogen atom, with one proton at the centre and one electron in orbit. The idea captured people's imagination, and cleverly explained why hydrogen absorbed and emitted only certain energies of light, as the electron could only occupy certain orbits.

SEARCH FOR THE NEUTRON

The planetary model was a good fit for hydrogen – one proton and one electron – but beyond that, the numbers failed to add up. The next element, helium, should have two protons and two lightweight electrons. But it actually weighed four times as much as hydrogen, not two. Lithium weighed seven times as much as hydrogen. What was going on?

The answer was that a third type of elementary particle was hiding in the atom. Ernest Rutherford thought that they would be some sort of neutrally charged combination of protons and electrons. He predicted that the particle would have similar mass to a proton but no charge, so it would not be slowed down by repulsion with electrons, and would therefore penetrate matter easily.

Rutherford asked his colleague James Chadwick to investigate these particles, which he called neutrons. Chadwick became very interested in an experiment by German physicists Walther Bothe and Herbert Becker. It seemed to be producing the kind of evidence that Rutherford had been looking for.

Bothe and Becker's experiment involved using polonium, a radioactive element that emitted alpha particles. When the pair bombarded a sample of beryllium metal with the particles, it yielded a powerful but unknown radiation, which Bothe and Becker thought might be gamma rays.

In France, Irène and Frédéric Joliot-Curie also investigated this radiation. They found it could dislodge protons out of paraffin wax, with the explanation

James Chadwick. ▶

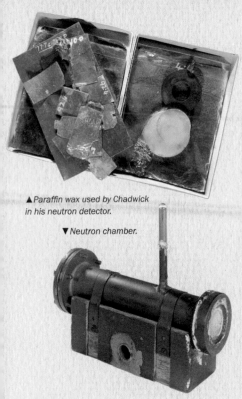

▲ *Paraffin wax used by Chadwick in his neutron detector.*

▼ *Neutron chamber.*

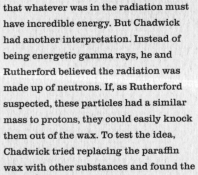

▲ *Irène and Frédéric Joliot-Curie.*

that whatever was in the radiation must have incredible energy. But Chadwick had another interpretation. Instead of being energetic gamma rays, he and Rutherford believed the radiation was made up of neutrons. If, as Rutherford suspected, these particles had a similar mass to protons, they could easily knock them out of the wax. To test the idea, Chadwick tried replacing the paraffin wax with other substances and found the experiment still produced protons. He announced the discovery of neutrons in 1932 – and that solved the mystery of the missing mass in Bohr's planetary model.

The helium nucleus, it turned out, contained two neutrons as well as two protons. For lithium, the number of neutrons was four. Throughout the periodic table, atoms have roughly the same number of neutrons in their nuclei as protons.

ISOTOPES

Nobel medal. ▶

Nobel medal. ▶

Neutrons helped explain a mystery about hydrogen itself. In 1931, American scientist Harold Urey had begun investigating whether there could be more than one type of the first element in the periodic table.

Previous research, carried out with radioactive elements, had shown the possibility that versions of the same element could exist, which had the same chemical properties but varying mass. For example, there were two versions of the element thorium, one produced by the radioactive decay of uranium, and the other by a different route from actinium. Chemist Frederick Soddy suggested in 1913 that these varieties of atom belonged in the same location in the periodic table, and called them *isotopes* after the Greek for "same place".

By the 1920s, isotopes of some light elements had been found, and Harold Urey wondered if hydrogen would be the same. He distilled liquid hydrogen and managed to find an isotope that was twice as heavy as normal hydrogen. But what was causing the extra weight? The answer came with the discovery of neutrons. Once again, everything fell into place. Isotopes of elements had the same number of protons and electrons, but different numbers of neutrons.

HYDROGEN'S DIFFERENT ISOTOPES

Hydrogen's two naturally occurring isotopes are extraordinarily rare. Deuterium, with a single neutron in its nucleus, makes up only 1 of every 6,400 hydrogen atoms in seawater. Tritium, with two neutrons, forms in trace amounts from the interaction of cosmic rays with the atmosphere and undergoes radioactive decay.

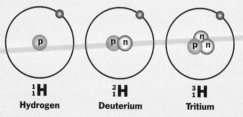

$^{1}_{1}\text{H}$
Hydrogen

$^{2}_{1}\text{H}$
Deuterium

$^{3}_{1}\text{H}$
Tritium

NOBEL PRIZE-WINNING ATOMIC RESEARCH

1906 Prize for Physics:
Electrons – J J Thomson
for "his theoretical and experimental investigations on the conduction of electricity by gases."

1908 Prize for Chemistry:
Atoms can change – Ernest Rutherford
for "his investigations into the disintegration of the elements, and the chemistry of radioactive substances."

1921 Prize for Chemistry:
Isotopes – Frederick Soddy
for "his contributions to our knowledge of the chemistry of radioactive substances, and his investigations into the origin and nature of isotopes."

1922 Prize for Physics:
Atomic structure – Niels Bohr
for "his services in the investigation of the structure of atoms and of the radiation emanating from them."

1935 Prize for Physics:
Neutrons – James Chadwick
for "the discovery of the neutron."

▲ *Frederick Soddy.*

Niels Bohr. ▶

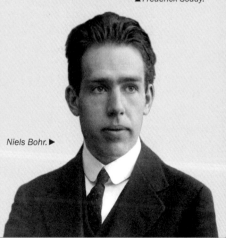

PERIODIC PIONEERS
ERNEST RUTHERFORD

Born into a family of 12 children, Ernest Rutherford began life on a rural farm in New Zealand, but travelled across the world to work in key scientific laboratories of his era, leaving a tremendous legacy.

Rutherford's parents were a flax-miller and a teacher who believed education and enquiry were key to problem-solving. Their influence was perhaps what led their son Ernest, who arrived in 1871, to gain a reputation as a straightforward and unpretentious leader in life and work.

Ernest did well at school and university and in 1895 gained a scholarship to the Cavendish Laboratory in Cambridge, UK. There, he worked with J J Thomson, who commented, "I have never had a student with more enthusiasm or ability for original research than Mr Rutherford."

A relentless spirit for discovery seemed to permeate Rutherford's work. On moving to McGill University in Montreal, Canada, in 1898, he collaborated with Frederick Soddy, who described working with his colleague as "hectic to a degree rare in the lifetime of an individual." Of the same period, Otto Hahn later recalled: "Rutherford's enthusiasm and abounding vigour naturally affected us all...he had a great, hearty laugh which echoed through the whole laboratory."

AN INSPIRING LEADER

Rutherford moved in 1907 to take up a physics professorship in Manchester, UK. Here, he performed his famous experiment to prove that the charge in an atom occupies only a tiny volume, the nucleus. Niels Bohr joined Rutherford and built on his ideas about the structure of the atom, saying, "Rutherford is a man you can rely on; he comes regularly and enquires how things are going and talks about the smallest details."

In 1919, Rutherford returned to the Cavendish Laboratory as director. During his period at the helm, James Chadwick discovered the neutron; John Cockcroft and Ernest Walton split the

▲ *John Cockcroft and Ernest Walton worked with Ernest Rutherford (centre) at the Cavendish Laboratory in Cambridge, UK.*

atom; and Edward Appleton proved the existence of the atmospheric ionosphere – all becoming Nobel Prize-winners. Rutherford's work had established the structure of the atom, as well as the nuclear process of radioactive decay.

Rutherford married a New Zealander called Mary Georgina Newton and they had a daughter, Mary. She married a physicist and astronomer, Ralph Fowler, and they had four children. In 2012, a great-grandchild of Rutherford, geophysicist Mary Fowler, became Master of Darwin College, Cambridge.

ELECTRONS AS WAVES

The model of the atom proposed by Niels Bohr in 1913 worked well for hydrogen, with a single electron that could jump up and down between discrete levels – and, given enough energy, even escape the atom altogether, leaving a hydrogen ion.

But in atoms with more than one electron, everything became more complicated. The model could not account for the interaction between electrons as well as with the nucleus. It did not explain the patterns of the periodic table, such as the fact that lithium, sodium and potassium on the left of the table are particularly reactive, or that helium, neon and argon, on the right, are inert.

Quantum theory stepped into the breach during the 1920s, with the work of Albert Einstein, Louis-Victor de Broglie and Werner Heisenberg. One major new idea was that electrons in the atom could be described more like

▲ Einstein as a young man.

▲ Potassium is a reactive metal on the left side of the periodic table.

ELECTRON CLOUD SHAPES

According to quantum theory, electrons do not actually move around the nucleus like little planets. Instead, they exist as probability clouds, in three-dimensional shapes called orbitals, denoted by letters s, p, d and f. The letters originate from an older system in which scientists classified spectral lines produced when an element was heated as sharp, principal, diffuse and fundamental.

Each orbital can accommodate up to two electrons – and, as the diagram shows, they come in different types. The s type has one orbital and can hold two electrons, while the p type has three orbitals, so can hold six electrons. The d type has five orbitals and can hold up to ten electrons, while the f type has seven orbitals.

waves than particles. In this model, there were still electron shells, like the previous energy levels. But the shells could be subdivided into mathematically defined subshells containing orbitals, which allowed a calculation of the probability of an electron being in a particular region around the nucleus.

While this system served for hydrogen, it also held for more complex elements. Whatever the element, its electrons existed in orbitals which took up certain shapes denoted as: s, p, d and f. The orbitals could also exist in many sizes, with more energy required for an electron to occupy larger orbitals.

ORBITALS AND SHELLS

As a rule, electrons would prefer to be in the lowest available energy state – so you might imagine they would all be taking it easy in the first of the s orbitals, closer to the nucleus. But work by Wolfgang Pauli on a quantum property of particles called spin showed that only two electrons could exist together in the same orbital. So as a result, the electrons fill an s orbital first (two electrons – the first shell), followed by an s and a p orbital (making up the second shell of eight electrons), then an s, a p and a d (the third shell of 18 electrons), and so on.

SHELL STRUCTURES

1 Hydrogen has a single electron in the s orbital of the first shell.

2 Helium has two electrons in the s orbital and thus a full first shell.

3 Lithium has a full first shell and a single electron in the second s orbital, which is in the second shell.

4 Neon has ten electrons, which gives it two full s orbitals and a full p orbital, meaning its second shell is full.

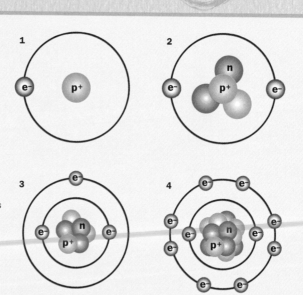

ORBITALS AND THE PERIODIC TABLE

The orbital model of electrons reveals why the periodic table works as it does. An atom that has a full outer shell tends to be nonreactive because a full shell is very stable. Elements from the last group of the periodic table, known as the noble gases, have full outer shells and are correspondingly inert.

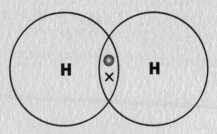

▲ Hydrogen atoms join together in pairs to share their electrons.

Hydrogen, on the other hand, has a single electron in its outer shell, like the alkali metals in the first group of the periodic table. These elements are all extremely reactive because it takes little energy for the electron to become involved in a chemical reaction and thus become more energetically stable.

Orbitals help to reveal how and why different elements bond together to form new compounds – the atoms are looking to achieve a more stable energetic state. For example, hydrogen atoms join together in pairs so that they can share electrons and have a full s orbital in their outer shell.

Even in this diatomic state, hydrogen is the lightest element of all. It is 14 times lighter than nitrogen, and 16 times lighter than oxygen – no wonder early pioneers of flight looked to hydrogen for a lift.

◀ Wolfgang Pauli.

LIGHTER THAN AIR

For centuries, the only way to rise heavenward was by climbing a nearby hill or tower to admire the view. During the 18th century, hydrogen gas played an important role in changing that.

In November 1783, a blue-and-gold balloon carried people into the sky over Paris. This was the first free flight in a hot-air balloon – although the pioneering Montgolfier brothers actually believed it was smoke that made the craft rise. They had seen the upward-flying ash of fires stoked in their family's paper-manufacturing business – and supplied old shoes and rotting meat for the balloon's brazier to keep the smoke billowing.

WHAT MAKES A BALLOON FLOAT?

The reason a hot-air or a hydrogen balloon floats in the air is because of buoyancy, a principle worked out by Archimedes, the ancient Greek mathematician. The balloon has a mass, and gravity pulls that mass downward to create the balloon's weight. But if the balloon is filled with a gas lighter than the air around it, it displaces a volume of air that weighs more than it does itself – so it floats.

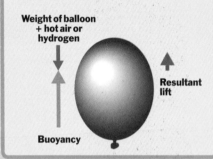

Weight of balloon + hot air or hydrogen

Buoyancy

Resultant lift

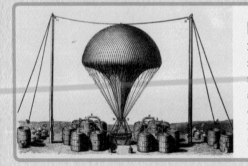

HYDROGEN HEROES

Edward Rudge published a recipe for making hydrogen using 19lb of iron filings, 40 gallons of concentrated sulfuric acid and five times the amount of water. He lived in Oxford, and may have supplied the gas for James Sadler's hydrogen-powered flight.

Less than a fortnight later came a second French first – a flight in a balloon filled with hydrogen gas. Inventor Jacques Charles and engineer Nicolas-Louis Robert launched their balloon from the Jardin des Tuileries, and flew for over two hours before landing safely.

Adventurers in France and Britain alike were gripped by balloon mania. In November 1784, a pastry chef called James Sadler made the first British hydrogen balloon flight, from the city of Oxford.

The benefits of hydrogen over hot air – despite the danger of its flammability – were twofold. First, the aeronauts could do away with the stove on board – reducing weight and also the tendency for sparks to set things alight. Second, because of the superior buoyancy of hydrogen, they needed less gas and could manage with a smaller envelope, as the balloon itself was known. For decades after, hydrogen was the favoured lifting gas for balloons.

▲ *The first human flight in a hydrogen balloon took place in 1783 over Paris.*

SAFE TO FLY

Even though hydrogen enjoyed buoyant popularity, it was the cause of several high-profile accidents. Among them was one that claimed the life of the first professional female balloonist, Sophie Blanchard. She fell from her craft when it caught fire over the Tivoli Gardens in Paris in 1819.

▲ *Sophie Blanchard, the first female professional balloonist, lost her life in an accident.*

Hydrogen also took the blame for the catastrophic demise of the German airship LZ 129 *Hindenburg* in 1937, during the last minutes of a flight from Frankfurt, Germany, to New Jersey, USA. While attempting to dock, it burst into flames, becoming "a puny plaything in the mighty grip of fate", as *Pathé News* described it, alongside fiery footage that cooled enthusiasm for hydrogen.

But fast-forward to today, and hydrogen is once again fuelling aviation. An experimental four-seater aircraft called the HY4 made its first official flight in 2016, powered by a hydrogen fuel cell system.

GREENER FLIGHT

Instead of burning jet fuel, this craft carries two high-pressure hydrogen tanks in its double fuselage, supplying four fuel cells in the central hull. They generate electricity to drive the propeller, taking the aircraft to a cruising speed of around 145kph (90mph). Also on board are two lithium polymer batteries to boost the plane's performance during take-off and when climbing.

Hydrogen fuel cells generate electricity via a chemical reaction between hydrogen and oxygen. The output is water, which is recycled naturally. The team building the HY4 at Germany's DLR Institute of Engineering Thermodynamics hopes that hydrogen-fuelled planes could serve as electric air taxis. For the project to be emission-free, however, the production of hydrogen from renewable sources will have to be scaled up considerably.

HOW DOES HYDROGEN FUEL A FUEL CELL?

Inside a fuel cell, there is a sandwich made of a positive electrode (the anode) and a negative electrode (cathode), with the middle made of polymer electrolyte membrane (PEM). Hydrogen flows past the anode, splitting into negatively charged electrons and positively charged protons. But only the protons can pass through the PEM layer, leaving the electrons as a useful electrical current that flows along a circuit. At the cathode, the electrons and protons recombine, along with oxygen from the air, to form water – the only waste product.

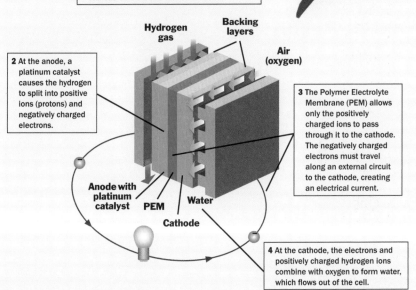

1 Hydrogen fuel is channelled through field flow plates to the anode on one side of the fuel cell, while oxygen from the air is channelled to the cathode on the other side of the cell.

Hydrogen gas

Backing layers

Air (oxygen)

2 At the anode, a platinum catalyst causes the hydrogen to split into positive ions (protons) and negatively charged electrons.

3 The Polymer Electrolyte Membrane (PEM) allows only the positively charged ions to pass through it to the cathode. The negatively charged electrons must travel along an external circuit to the cathode, creating an electrical current.

Anode with platinum catalyst

PEM

Water

Cathode

4 At the cathode, the electrons and positively charged hydrogen ions combine with oxygen to form water, which flows out of the cell.

3	
Li	
Lithium	

11	
Na	
Sodium	

19	
K	
Potassium	

37	
Rb	
Rubidium	

55	
Cs	
Caesium	

87	
Fr	
Francium	

CHAPTER 3

ALKALI METALS

How did the Bunsen burner transform
the quest for new elements?

Why are substances in this group vital for thought?

What is driving the search for lithium in our world?

FAMILY TIES

What gives your family its character? The alkali metals are a family that share distinct physical characteristics, in particular a silvery appearance and soft texture in pure form. As a clan, however, they have a bit of a reactive reputation. Many people know them for their tendency to respond violently, particularly to situations involving water, and even bursting into flames when exposed to oxygen in the air.

▲ *Pure sodium reacts rapidly in the air if not stored in oil.*

It is the harsh chemicals produced when these elements come into contact with water that give the group their name – the alkali metals. For example, sodium and water react to produce hydrogen gas and sodium hydroxide, which also goes by the name of caustic soda or lye. One of the most powerful alkalis, it is extremely corrosive but it is vital for industrial soap- and paper-making, and makes an effective drain cleaner when nothing else will work.

The alkali metals do not exist in pure form in the natural world because they are so reactive. For experiments in the school laboratory, chemistry teachers have to keep samples of pure lithium in liquid paraffin or oil, and pure sodium in a jar of mineral oil, to stop them reacting with water vapour and oxygen in the air.

What is the chemical reason for this? In terms of their atomic structure, all the family members have a single outer electron, which makes them very reactive. That lone outer electron wants to leave, and takes every opportunity to do so, turning the metal into a positively charged ion. If there is something else around, like a chlorine atom, the electron joins that instead, filling a suitable gap, and giving it an overall negative charge. The positive ion attracts the negative, and a strong ionic bond forms between sodium and chlorine ions in an ionic crystal lattice: NaCl or table salt.

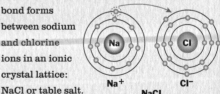

▲ *Sodium (Na) bonds with chlorine (Cl) to form table salt.*

THE pH SCALE

	pH	Example
Strong alkali	14	Drain cleaner
	13	Bleach
	12	Soapy water
	11	Ammonia
	10	Milk of magnesia
	9	Baking powder
	8	Seawater
Neutral	7	Pure water
	6	Saliva
	5	Black coffee
	4	Acid rain
	3	Vinegar
	2	Lemon juice
Strong acid	1	Stomach acid
	0	Battery acid

Acids and alkalis sit along the pH scale, with battery acid at one end of the range, and alkalis such as drain cleaner at the other. Water is a neutral party in the middle, with a pH of 7. An acid and an alkali neutralize each other.

71

Fast Facts
LITHIUM
ATOMIC NUMBER 3

▼ *Pure lithium.*

Character: Lithium is the world's lightest metal, silvery-white in colour when pure and the least dense of all the solid elements.

Discovered: Swedish chemist Johan August Arfwedson discovered lithium in 1817 in a chunk of rock called petalite. He announced the find in 1818, and correctly identified the new substance as an alkali metal, but was unable to separate enough to perform experiments.

Name: Lithium had a starry start, as one of the three elements created during the Big Bang, along with hydrogen and helium. Despite this, lithium's discoverer named it after the Greek for stone, *lithos,* because of the substance's occurrence in rocky minerals.

World sources: Today, most lithium production is via evaporated brine from salt flats in Bolivia and Chile, while some is mined from deposits in the USA.

▲ *Petalite yielded the first sample of lithium in 1817.*

Bolivian salt flats. ▶

▲ *Lithium metal reacts violently with water.*

JUST ADD WATER

Table salt is certainly useful, but it is not spectacular, like the pure alkali metals. If they are dropped into water (using suitable safety equipment and precautions – these reactions are very lively), this series of metals reacts with increasing vigour, all giving off hydrogen gas. Lithium and sodium float on the water and dance around as they fizz. Sodium can catch alight as it skips about, but potassium is even more impressive, bursting into flames as the heat of the reaction ignites the flammable gas. Rubidium is gone with an energetic bang as soon as it hits the water. Caesium gives out so much energy, so fast, the explosion may take your experimental equipment with it.

Elements in this group react more violently as you go further down the group because the atoms are larger. This means the outer electron is held less tightly by its own nucleus and becomes even keener to leave.

Francium should follow this pattern, but it is extremely unstable and radioactive, and very short-lived, so you cannot easily do experiments with it at all. However, the theory goes that because francium is such a heavy element with so many protons in its nucleus, its electrons end up more difficult to remove than expected. If we could get any francium to test, therefore, it would be slightly less reactive than predicted by the family pattern.

With all the talk of harsh alkalis, explosions and radiation, you might have thought that this group of elements would be best avoided for your health. But the reality is that, in the right compound, many of these chemicals play vital roles in our lives.

▲ *Rechargeable battery manufacturing uses vast quantities of lithium.*

LITHIUM FROM LAKES

In southwest Bolivia is the world's largest salt flat, the Salar de Uyuni. It is a desert nearly 11,000 sq km (4,000 sq miles) in size. A few centimetres beneath its crust is a precious blue-green salty brine, which produces the major proportion of the world's lithium.

The Bolivian government is busy extracting the metal using large evaporation ponds, and processing it for sale. The resulting lithium carbonate is snapped up by

▼ *The Salar de Uyuni, a huge salt flat in Bolivia, is a key lithium source.*

LITHIUM-ION BATTERIES

Power-packed lithium-ion batteries contain a positive electrode made of lithium, and a negative electrode made of carbon, immersed in a solvent and kept under pressure. When the battery is in use, lithium ions flow from one electrode to the other, while useful electrical energy flows out of the battery to run your gadget.

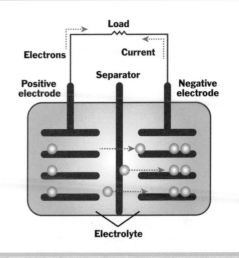

Lithium-ion rechargeable battery discharge mechanism. ▶

manufacturers to feed a global appetite for lithium-ion rechargeable batteries in portable devices such as laptops and mobile phones.

Nobody could have foreseen such a lithium boom when lithium-rich minerals were first found, in an iron mine on the Swedish island of Uto, in the 1790s. Named petalite and spodumene, large deposits of these minerals exist in Australia, Brazil, Namibia and the USA, where a single crystal of spodumene was unearthed in South Dakota weighing over 10 tonnes.

But extraction from brine is now growing rapidly, as producers scramble to meet new demands for lithium in electric car batteries, alongside other gadgets.

New research is also suggesting that a solid glass-based electrolyte could help solve some of the problems of safety, energy density and recharge speed associated with current lithium-ion batteries. Made using sodium instead of lithium, the new battery could treble energy storage, researchers claim. It can also operate at -20°C (-4°F), while earlier solid-state batteries required a sizzling 60°C (140°F) to function. But will it really work? Electric vehicle fans await news with interest.

Fast Facts
SODIUM Na
ATOMIC NUMBER 11

Character: Sodium is a soft, silvery-white metal. Like its closest relatives in the alkali metal family, a cut surface is shiny only for a few seconds before it reacts with oxygen in the air and becomes dull.

Discovery: In 1807, in Britain, Humphry Davy extracted sodium metal from molten sodium hydroxide. The new process of electrolysis yielded metallic globules of sodium.

▲ *Pure sodium.*

Name: Soda is an age-old name for substances useful for cleaning and baking, and Humphry Davy coined sodium as the name for the metal he isolated from caustic soda. But its chemical symbol Na comes originally from the Natron Valley in Egypt, where ancient people used soda for washing, soaps and even in mummification.

World sources: Abundant around the world, sodium metal is produced from molten sodium chloride: common salt.

POTASSIUM PERIL

Are bananas radioactive?▼

Bananas are full of good things like magnesium, iron, and lots of potassium. It sounds alarming to realize that a proportion is the radioactive potassium-40. Do not worry too much, though – you would have to eat 1,052,417,518,000 bananas at once to receive a lethal dose. And long before that time came, your body would have eliminated the same amount of potassium as it had taken in, to retain its own natural levels – a process called homeostasis.

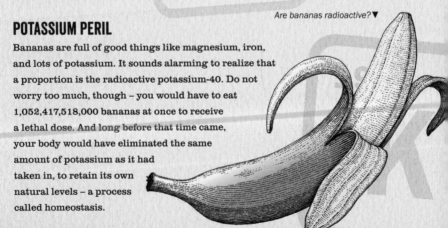

Fast Facts
POTASSIUM K
ATOMIC NUMBER 19

Character: A soft, silver metal, potassium is light enough to float on water. As it reacts, it releases hydrogen gas which ignites, and the metal also burns with sparks and a purple flame.

Discovery: British chemist Humphry Davy discovered potassium in 1807, going on to find several alkaline earth elements including magnesium, calcium and strontium.

Name: When Humphry Davy first isolated this new metal, he named it potassium after the caustic potash – potassium hydroxide – from which he had extracted it. Potash is what remains from burned wood ashes, soaked in water, and it has been a useful substance for centuries in the textile industry, in glass- and soap-making, and even for preparing gunpowder. Its symbol K comes from *kalium*, the word for potash in Latin.

World sources: Today, potassium is mined in Canada, Russia and Belarus, from minerals deposited in lake and sea beds millions of years ago, and often now deep underground.

◄ *Pure potassium floats on water, giving off flammable hydrogen gas.*

▼ *Potash evaporation ponds in Colorado Plateau, Utah, USA.*

SODIUM AND POTASSIUM IN YOUR BODY

Among the elements in your body, those making up the largest proportion are oxygen (65 per cent), carbon (18.5 per cent) and hydrogen (9.5 per cent). Since we are a carbon-based species, with a huge amount of water inside us, this is no surprise.

Focussing on smaller percentages, however, we find potassium (0.4 per cent) and sodium (0.2 per cent). So what are these alkali metals doing in the body? They perform a vital role within neurons, a key building block of your nervous system, which handles every thought and sensation you ever have. Nerve impulses are electrical signals which the nerve creates using sodium and potassium ions – atoms that have lost their outer electron and become positively charged.

All your nerve cells work by moving sodium and potassium ions around. Even at rest, a nerve cell is busy pumping sodium ions out of itself, and pumping potassium ions in. Both kinds of ion tend to flow back to their original

Potassium, sodium and other elements	Element	Symbol	% in Body
Nitrogen — 3.2%	Oxygen	O	65.0
	Carbon	C	18.5
Hydrogen — 9.5%	Hydrogen	H	9.5
	Nitrogen	N	3.2
	Calcium	Ca	1.5
	Phosphorus	P	1.0
Carbon — 18.5%	Potassium	K	0.4
	Sulfur	S	0.3
65%	Sodium	Na	0.2
	Chlorine	Cl	0.2
	Magnesium	Mg	0.1
Oxygen —	Trace elements include boron (B), chromium (Cr), cobalt (Co), copper (Cu), fluorine (F), iodine (I), iron (Fe), manganese (Mn), molybdenum (Mo), selenium (Se), silicon (Si), tin (Sn), vanadium (V) and zinc (Zn).		less than 1.0

SODIUM SIGNALS

You do not want to meet a golden dart frog if it is feeling grumpy. When alarmed, it can release a substance from its back and behind its ears called batrachotoxin. This neurotoxin works by forcing open all the channels in your nerves that carry sodium. The nerves can no longer send signals, so your muscles are paralyzed.

▼ *The largest species of golden dart frog, this animal is the most toxic in the Amazon rainforest in Colombia.*

locations, but the potassium manages this more quickly, so the cell ends up with a negative charge inside. This is called the resting potential.

When a receptor stimulates the cell, it is action stations for the ions. Channels in the cell surface open, which means positive sodium ions can flood in from outside. They start to reduce the negative charge inside the cell, as you would expect. When the charge has reached a positive threshold level, suddenly the cell fires and transmits the signal.

After that, the cell closes to sodium, and the potassium channels open, so that its ions can flood out of the cell and reduce the overall charge inside. Finally, with all the channels closed, the pumping begins again to set up the negative charge inside.

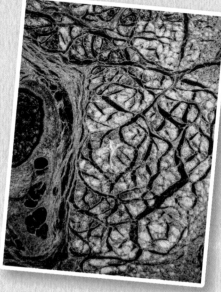

▲ *Nerve cells all over your body rely on sodium and potassium ions to transmit signals.*

DESALINATING SEAWATER

More than one in six of the world's people face water scarcity. Desalination – removing salt from seawater to make it drinkable – is one energy-hungry and expensive solution.

Industrial desalinization plants usually boil seawater to separate salt, or filter it through polymer membranes. Scientists at the University of Manchester, UK, have managed to fashion graphene-based sieves to separate out the salt.

The team used graphene oxide, which they can produce easily in the lab, to create membranes with pores so small that they prevent sodium chloride passing through. The membranes allow a fast flow rate, and the scientists hope it will help create desalination devices that work with low energy input.

▼ *Reverse osmosis uses a semipermeable membrane to remove unwanted ions and molecules from water to clean it for drinking.*

SPOTLIGHT ON RUBIDIUM AND CAESIUM

With their trusty flames and flexible rubber tubes, Bunsen burners are a fixture in chemistry laboratories all over the world. They were the bright idea of Robert Bunsen, who asked his university's instrument-maker, Peter Desaga, to construct the new burners in 1855 – and they played an important role in the discovery of two elements among the alkali metal group.

Bunsen was a chemistry professor at the University of Heidelberg, Germany, and his career came at a time when laboratories were becoming more and more sophisticated. Earlier researchers had tended to carry out their experiments in rather basic workshops. But the lab Bunsen set up at Heidelberg laid a pattern for those who came after, with glass-fronted cupboards for supplies, preparation rooms, a library and long benches. Bunsen oversaw these developments – and even had a specialized fume hood installed for smelly experiments, called the *Stinkzimmer*.

German chemists Gustav Kirchhoff (left) and Robert Bunsen. ▶

At Bunsen's direction, the lab had a mains water supply and was also connected to the gasworks that opened in the city in 1854. This paved the way for the burner to replace the fiddly alcohol and oil lamps everyone had previously contended with. It mixed gas and air to give a small, hot, steady non-luminous flame, ideal for the modern lab.

The new burner led Bunsen and a colleague from the university's physics department, Gustav Kirchhoff, to develop skills in a key chemistry technique – spectroscopy – and uncover a host of new elements.

COLOURFUL BARCODES

Alchemists and chemists alike had known for centuries that substances produce characteristic colours if you burn them. By 1800, eight specific colours had been linked to elements – boron, sodium, copper, potassium, magnesium, calcium, strontium and barium. Chemists of the period would perform blowpipe tests on a new substance, in which they puffed a quantity into a flame to see if it burned with a specific colour.

A problem with this technique, however, was that the smallest trace of sodium would overwhelm any other colours with yellow. Blowpipe analysis was limited by this, as well as the difficulty experimenters found in creating a hot and reliable flame.

Bunsen and Kirchhoff set out to examine not just the colour of a substance in the flame, but its emission spectrum when split using a prism. Kirchhoff provided apparatus from the physics lab to set up a spectroscope, allowing the pair to map the specific pattern of

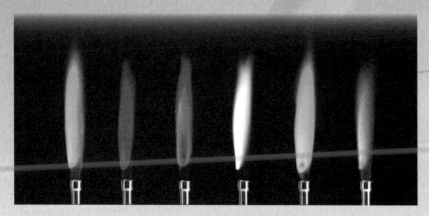

▲ *Alkali metals are among the elements responsible for the colours in these flame tests. From left to right: barium (pale green), strontium (red), lithium (red), sodium (yellow), copper (green) and potassium (lilac).*

wavelengths that appeared when they heated a substance in the Bunsen flame. It was like being able to read a barcode of the elements within the substance, no matter how large or small the quantity present, and no one had done it in such a systematic way before.

Bunsen was keen to look for new elements everywhere. In 1860, he and Kirchhoff analyzed a sample of mineral water from the nearby spa town of Durkheim, and noted a pair of blue lines they did not recognize. After commissioning a local chemical factory to evaporate 55,000 litres (12,000 gallons) of the water, Bunsen received a mighty yield of 7g (0.25oz) of a compound. It would be enough to ascertain that it contained a new metal – caesium.

Bunsen and Kirchhoff also experimented with a mineral called lepidolite. They knew it contained lithium and potassium, but having removed their telltale lines from the spectroscopic results, the two researchers found it held a secret. The two deep red lines that remained showed they had found another new element – rubidium.

BUNSEN AND KIRCHHOFF'S SPECTROSCOPE

The sample holder held a substance in the flame of the Bunsen burner and a telescope focussed the light into a box containing a prism, which split the light into a spectrum that could be viewed through a second telescope – as this historic image shows.

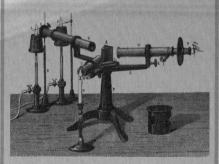

◀ Lepidolite is an ore of lithium and also one of the major sources of the rare alkali metals rubidium and caesium.

EMISSION SPECTRA FOR THE ALKALI METALS

By putting a compound into a hot flame, electrons inside the atoms are temporarily excited to higher energy levels. When they drop back down to a lower level or ground state, they release energy as light at characteristic wavelengths. The spectroscope splits the light using a prism to reveal the spectrum as a series of sharp lines.

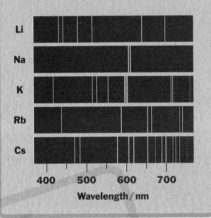

Fast Facts
CAESIUM Cs
ATOMIC NUMBER 55

Character: Caesium is a pretty, gold-coloured liquid when just above room temperature.

Discovery: The first element to be identified by the process of spectroscopy, caesium was found in 1860 by Gustav Kirchhoff and Robert Bunsen. Carl Theodor Setterberg went on to isolate the pure metal, an achievement he published in 1881.

Name: Bunsen and Kirchhoff named the element after the Latin word for sky blue, *caesius*, because of the blue flames and spectroscopic lines it produced.

World sources: Caesium is extracted from the mineral pollucite, which is mined in Canada and Zimbabwe – but is mostly a by-product of the lighter alkali metal lithium.

▲ *Pollucite is a significant source of caesium and sometimes rubidium.*

▲ *Caesium clock vacuum chamber.*

RUBIDIUM AND CAESIUM CLOCKS

You may never hear it tick or tock, but rubidium turns a rock into a clock. All minerals contain radioactive elements such as rubidium, which act like independent timekeepers. As rubidium decays, it forms the more stable daughter element strontium. The relative proportions of each tell us how long the rock has been around.

Caesium atoms are the basis of the most accurate clocks, which enable us to run GPS systems. Atomic clocks keep time based on the frequency of electromagnetic radiation released when electrons transition between energy states in a caesium atom – and they will not lose or gain even a second in 158 million years.

▶ *Caesium melts at just above room temperature.*

FINDING FRANCIUM

At the age of only 19, in 1929, Marguerite Perey applied for a job at the prestigious Radium Institute in Paris. Her interviewer was Marie Curie, a huge scientific star who had won two Nobel Prizes, developed the theory of radioactivity and studied its effects, applying them in radiotherapy for the treatment of cancers.

Perey became Curie's lab assistant, and embarked on a daily pursuit of the radioactive element actinium for Curie to study. Research continued at the Institute after Marie Curie died in 1934, and Perey began to notice that the actinium gave off unexpected radiation. She managed to purify a sample of the actinium compound to study the emissions, and realized that while nearly all the actinium decayed to form thorium, about 1 per cent decayed instead to form a new element.

THE MISSING ELEMENT

Plenty of claims had previously been made relating to an element to fill a gap at the bottom of the alkali metals. Perey deduced she had found the missing

◄ *The Radium Institute, Paris, where Marie Curie was director of research from 1918 until her death in 1934.*

▲ *Marguerite Perey in 1962.*

element 87, produced as the actinium underwent alpha decay. The particles quickly decayed again to radium.

Perey gained her Ph.D with a thesis about the element she called francium, and became the first woman admitted to the Academie des Sciences in France. But in 1946 she noticed skin burns from prolonged exposure to radiation from the actinium, and died from bone cancer in 1975.

Fast Facts
FRANCIUM [Fr]
ATOMIC NUMBER 87

Character: Francium is radioactive with no stable form, and lasts for only about 20 minutes.

Discovery: Marguerite Perey discovered francium in 1939. Many people had searched for an element to fill the gap at position 87 in the periodic table, but it was not until Perey sought the new element in the decay products of other radioactive compounds that she spotted the elusive substance.

Name: Catium was the first name Marguerite Perey suggested, in recognition that this element would be keen to lose its single outer electron and become a positively charged ion – a cation. However, eventually she chose francium after her native France.

World sources: At any one time, less than 30g (1.05oz) of francium exists naturally via the decay of other radioactive elements. To make francium artificially, you have to bombard radium with neutrons in a nuclear reactor.

◄ *In thorium minerals, a few atoms of incredibly rare francium exist through radioactive decay.*

4
Be
Beryllium

12
Mg
Magnesium

20
Ca
Calcium

38
Sr
Strontium

56
Ba
Barium

88
Ra
Radium

CHAPTER 4

ALKALINE EARTH METALS

**WHICH OF THE ALKALINE EARTH METALS
ACCOUNTS FOR A KILOGRAM OF YOUR WEIGHT?**

**WHY DO WE NEED TO FIND A GREEN SOURCE
FOR GREEN FIREWORKS?**

WHAT MAKES MARIE CURIE SUCH A SCIENTIFIC STAR?

ALKALINE EARTH METALS

The alkaline earth metals are the periodic table's second vertical group of elements. They are similar in many ways to the alkali metals group next door, but slightly less reactive – rather like calmer, weightier siblings. Nonetheless, they are responsible for many of the exciting colours in firework displays.

Their name recalls that they were all originally found in very earthy sources – limestone mortar, pretty minerals and hefty rocks from lead mines. But this group caused great frustration to the famous chemist Antoine Lavoisier.

He characterized the substances he called earths as being insoluble in water and resistant to heat. Before he died in 1794, however, Lavoisier expressed a suspicion that, far from being elemental, the earths were in fact compounds of metals with oxygen. He turned out to be right. In 1808, by the power of the newly invented battery, Humphry Davy successfully separated a whole series

of shiny metals from their oxides and revealed magnesium, calcium, strontium and barium.

KEY BREAKTHROUGHS
Ever since, the alkaline earth metals have been involved in key aspects of chemistry's development. Davy performed his metal-producing experiments using electrolysis, a process named and perfected later by his eminent assistant, Michael Faraday. Electrolysis had also yielded potassium and sodium,

▲ Chrysoberyl is a beryllium-yielding mineral. This sample is green chrysoberyl.

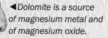

◀ Dolomite is a source of magnesium metal and of magnesium oxide.

▲ In a limekiln, limestone burns to produce quicklime – calcium oxide – a compound of one of the alkaline earth metals.

◄ Aragonite is one of the most common naturally occurring crystal forms of calcium carbonate.

▶ Emerald gemstones are a variety of beryl, containing beryllium. The green colour comes from traces of chromium.

and would later reveal lithium, gallium and fluorine to the world. Today, electrochemistry – the interaction between electrical energy and chemical change – has applications from the largest industrial power plants right down to the scale of nanotechnology.

Topping and tailing the group are elements with significant links to fundamental atomic discoveries. Down at the bottom, the heaviest of the alkaline earths is radium, an emitter of alpha radiation. Ernest Rutherford used it in 1909 in the famous experiment in which he fired particles at a piece of gold foil – and found that some of them bounced back (see page 53). It showed that there was a tiny, dense nucleus in the middle of the atom.

The lightest of the alkaline earth metals, beryllium, gave James Chadwick the clue in 1932 that neutrons were lurking in every atom. He studied the penetrating radiation produced when a beryllium sample was bombarded with alpha particles, and showed it must consist of neutrally charged particles, since they could knock protons clean out of other materials such as wax.

HI-TECH BERYLLIUM

All the alkaline earth metals have two electrons in their outermost shell. But beryllium is a bit of an oddity. It has a high melting and boiling point compared with its relatives, and because of its small atomic radius, it gives up its outer electrons less readily than others in the same group.

Nonreactive with water or air, light and strong; it sounds like a perfect metal for manufacturing – and it is. However, the difficulty is that beryllium compounds are so toxic in powder form that a lung disease is named after it: berylliosis.

That did not stop the motor-racing industry taking a spin with beryllium in the late 1990s. Some of the leading teams fitted pistons in their racing cars made with lightweight, stiff beryllium mixed with aluminium. It would no doubt boost their drivers'

▲ Emerald rocks contain beryllium, which is a bit of a scientific mystery.

◄ Beryllium is the perfect element from which to make a low distorting instrument like the Spitzer Space Telescope.

chances. But the governing body for world motor sport, the FIA, grew concerned both about the cost of the metal which excluded smaller constructors, and about the risks posed by beryllium dust. In 2001, they banned the element – despite protests from its fans.

However, in space technology, cost and safety calculations work differently, and beryllium has found particularly cool applications there. NASA's Spitzer Space Telescope, launched in 2003, observes incredibly distant stars by detecting faint heat radiation. Any distortion in the equipment will interfere with its view –

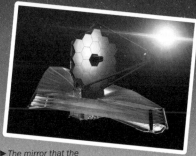

▶ The mirror that the James Webb space telescope carries is made of beryllium because it resists distortion.

so to keep readings precise, Spitzer is made almost entirely of beryllium, which has very low heat capacity. The structure of the telescope responds uniformly to any temperature fluctuations.

Fast Facts
BERYLLIUM Be
ATOMIC NUMBER 4

Character: Beryllium is a greyish-silver metal with a melting point much higher than all the other elements in its group. It is transparent to X-rays.

Discovery: The French mineralogist René-Just Haüy asked a colleague called Nicolas-Louis Vauquelin to investigate the gemstones beryl and emerald, suspecting they contained a novel element. In 1798, Vauquelin announced his discovery of the new metal, although it took until 1828 to isolate it.

Name: Beryllium's original name was proposed as glaucinium from the Greek for sweet, because its compounds apparently had a sweet taste. However, the alternative name stuck, based on the Greek name for the gemstone, beryl.

World sources: Very little beryllium is produced each year, because it is highly toxic.

METALLIC MAGIC

If you could have attended one of Humphry Davy's lectures at the Royal Institution in London, you might have seen frogs' legs twitching by the power of electricity, or a metal nugget bursting into flames after being thrown into water. The 23-year-old joined the august organization in 1801 and his effervescent chemistry talks soon became popular society events.

To assist with his research and demonstrations, Davy installed a device recently developed by Italian scientist Alessandro Volta – the voltaic pile, or battery. It offered Davy the chance to perform the process we now call electrolysis – passing a direct current through a substance to try to make it decompose.

▲ The voltaic pile.

◄ A young Humphry Davy assists at a chemistry lecture in this early 19th-century caricature of a lively event at the Royal Institution.

In 1807, Davy applied the battery to molten potassium hydroxide, a substance known for centuries as potash. As the current flowed through positive and negative electrodes made of platinum, Davy was thrilled to see droplets of a shiny metal appear from the melted substance. When he tested it, the metal revealed itself as reactive potassium. Davy managed to repeat his success with sodium a few days later.

The following year, Davy's experiments focussed on the alkaline earth metals. Progress was difficult until some correspondence and advice from Swedish chemist Jacob Berzelius, almost his exact contemporary, whose interest had also been piqued by electrochemistry. He suggested using mercury as the negative electrode instead of platinum. In this way, for the first time, Davy used electrolysis to create amalgams of magnesium, calcium, strontium and barium, and then removed the mercury from these new metals. Through Davy and Berzelius's work, samples of these reactive metals had been created, free from oxygen, for the first time – a spectacular success.

Character: Magnesium is a shiny, grey, lightweight metal. Take care, though: it burns with a bright white light in air, emitting ultraviolet radiation that can burn your retina causing scarring and even blindness.

Discovery: In 1755, Scottish experimental chemist Joseph Black first proposed that magnesium was an element. Humphry Davy separated the metal in 1808.

Name: Magnesium comes from the name of the Greek region Magnesia, after a mineral that yielded the metal. Humphry Davy proposed an alternative name, magnium, to avoid confusion with manganese, but magnesium stuck.

World sources: The eighth most common element in the Earth's crust, magnesium is abundant in the oceans, and most magnesium manufacture is from evaporated seawater.

CONVERSATIONS ON CHEMISTRY

Michael Faraday had a tremendous impact on both chemistry and physics, formulating the laws of electrolysis and discovering the principles of electromagnetism. But the story of how he happened to leave such a legacy comes back to the power of books.

Born into a poor family, Faraday went out to work at the age of 14 and became apprenticed to a bookbinder in London. He served there for seven years, taking the opportunity to read many of the texts that passed through his hands.

One of the titles that captivated him was a volume called *Conversations on Chemistry,* by Jane Marcet. She had been attending lectures by Humphry

▲ *Michael Faraday.*

◄ *Jane Marcet, author of* Conversations on Chemistry.

Davy at the Royal Institution, repeating experiments at home, and writing simple explanations in the form of a discussion between three characters – Mrs B, and two children, Caroline and Emily. Faraday, too, began trying the chemistry experiments for himself.

When Faraday had the chance to go to hear Davy lecture in person, he took copious notes, bound them into a book, and sent them to Davy. As a result, Faraday ended up working at the Royal Institution for 54 years, first as Davy's assistant and later as a professor of chemistry.

MAGNESIUM PROVIDES PRICELESS EVIDENCE

Diamonds from Brazil have revealed a clear-cut clue that there could be an ocean a kilometre deep in the Earth. The 90-million-year-old gems originally emerged from a volcano in Juina, western Brazil.

▲ An imperfection in a diamond from Brazil contained clues of an underground ocean.

Scientists investigated a curious imperfection in their structure – an inclusion of material trapped when the diamond formed.

The inclusion contained a mineral composed of iron and magnesium oxide combined with a range of other elements that showed the diamond was made deep down in the Earth's mantle.

Tantalizingly, the inclusion also provided evidence of the presence of water – hydroxyl ions.

It may seem very strange to think of an underground ocean. However, the researchers think that water plays a huge role in the process of plate tectonics on Earth, weakening or promoting melting in rocks and enabling them to move.

▼ Formed because of collisions and movements of tectonic plates, the Ring of Fire around the Pacific Ocean has over 400 volcanoes.

LIVING WITH MAGNESIUM AND CALCIUM

While the first of the alkaline earth metals could kill you, the second and third you cannot live without. Magnesium is vital to the function of hundreds of enzymes in the human body – biological catalysts that, among many other things, regulate glucose levels and enable DNA to replicate. Fortunately, human bodies store about 20mg of magnesium, mostly in the bones, and a varied diet usually supplies the 250–350mg intake required each day.

For plants, magnesium also plays a central role. In the middle of each molecule of the green pigment chlorophyll is a magnesium ion, surrounded by a ring of other molecules. The magnesium kicks off the process by which the plant captures sunlight and converts carbon dioxide and water into glucose, giving us oxygen as a handy by-product.

Calcium, too, is essential to life for everyone, other than for a very few insects and bacteria. Right now, you have about a kilogram (2.2lb) of calcium in your body – most of it in your bones and teeth, where it performs a vital structural function in the form of calcium phosphate. A few grams play other roles, making muscles contract, helping conduct nerve impulses, triggering hormones to be released, and controlling cell division. In the animal kingdom, calcium carbonate is the basis of eggshells, snail shells and seashells alike, while in plants it is an essential nutrient for building cells and communicating between them. It is truly crucial stuff.

Calcium is vital in bones, teeth and shells. ▲

CHLOROPHYLL

Every molecule of the plant pigment chlorophyll contains a magnesium ion.

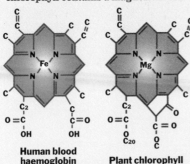

Human blood haemoglobin

Plant chlorophyll

▼ *Chalk cliffs formed from calcium-rich shells of marine creatures.*

CALCIUM Ca
ATOMIC NUMBER 20

Character: Calcium metal is a silvery-grey solid.

Discovery: Metallic calcium was one of several elements that Humphry Davy isolated in 1808 using the new technique of electrolysis. French chemist Antoine Lavoisier had previously suspected that a substance he knew as chaux – calcium oxide – might contain a new element.

Name: Calcium's name comes from the Latin word *calx*, meaning lime, because of its link with the building material lime used in making mortar, calcium oxide.

World sources: Many countries have deposits of calcium ores such as calcite, dolomite and gypsum, which are useful for producing lime and plaster for the construction industry. Calcium metal is produced by China, Russia and the USA for steelmaking and battery manufacture.

PRETTY PYROTECHNICS

A sparkling silver comet, a crimson chrysanthemum, a purple peony, a golden star shell…fireworks offer a feast of colour and light. Many of us around the world celebrate special events with impressive firework displays, which draw on many of the alkaline earth metals to create their bursts of crowd-pleasing colour.

To propel a firework into the air, you need black powder: a mixture of potassium nitrate, charcoal and sulfur. Once ignited with a spark or flame, the powder deflagrates, meaning that it burns relatively slowly. The potassium nitrate decomposes, releasing oxygen gas, which in turn increases the rate of reaction with the fuel, elevating the temperature. A lot of hot gases are produced to force the firework upward, as well as dirty smoke.

Packed into the firework is the payload – more black powder (the bursting charge), plus small black balls called stars, which hold the secret of the firework's sound and colour. At the right moment, the time-delay fuse ignites the bursting charge, which explodes, sending the stars, burning, out in every direction with a bang. The stars contain metal salts, and the intense heat excites the electrons in the metal atoms, which

▲ In 1685, fireworks on the River Thames in London marked the coronation of King James II and his Queen consort Mary of Modena.

then emit light of characteristic colours as they fall back to their more stable energetic states.

Red strontium carbonate, lithium carbonate

Orange calcium chloride

Yellow sodium chloride

Green barium chloride

Blue copper chloride

Purple mixture of strontium and copper compounds

Silver magnesium metal or aluminium (or the alloy magnalium)

Strontium sample. ▶

Character: Another soft, silvery metal, strontium is highly reactive, and turns yellow in the air as the surface oxidizes.

Discovery: A Scottish doctor and chemist called Adair Crawford first spotted the mineral that contained strontium in 1787. An Edinburgh mineral dealer offered the specimen as a possible source of barium, which interested Crawford for medical purposes. However, in 1790 he published his finding of the new and different element.

Name: Strontium is named after Strontian, a village on the rocky west coast of Scotland, whose local lead mine first yielded the strontium mineral, strontianite.

World sources: China, Spain, Mexico and Argentina produce strontium from deposits of the strontium mineral celestine.

◀ Black powder was the original gunpowder invented in China as long ago as 2,000 years.

▲ Celestine is a mineral that yields strontium.

GREENER CHEMISTRY

Popular though they are, firework displays have been causing environmental headaches. Some scientists are concerned about the contribution of perchlorates to water pollution. Perchlorates may disrupt the thyroid gland, affecting children's development. There is another worry over the release of barium into the environment, causing bronchial problems, and building up in the food chain.

As a result, chemists have been on the lookout for new chemicals that might replace the combustion-boosting perchlorates in fireworks. Nitrogen-rich energetic materials, such as tetrazole and tetrazine and their derivatives, are showing promise, although currently at a higher price. They store energy in bonds between nitrogen atoms, or between nitrogen and hydrogen atoms. When the combustion reaction takes place, stable products such as nitrogen gas are produced, releasing vast amounts of energy in the process. And they burn without producing a lot of smoke – reducing air pollution, but also allowing the firework colour to shine brighter while using less metal salt.

That still leaves the problem of the barium – far from being an environmental option. Instead, the scientists may be able to make a complex of the nitrogen-rich tetrazole with a metal such as copper to try to make greener greens.

STRONTIUM SKELETONS

Blink and you will miss them. Acantharea are minute single-cell organisms that live in the ocean and build their skeletons from strontium sulfate that they extract from the water. Long spikes called spicules radiate symmetrically in their complex layouts, each one a single crystal structure.

Acantharea seem incredible – but have been studied less than you might expect because it is so hard to preserve their skeletons after death. In the usual fixatives, their intricate structures rapidly dissolve and are never seen again.

BARIUM AND THE BOLOGNA STONE

The rules of alchemy say that the first step toward transforming a material to gold is to heat it. Thus, when a budding alchemist unearthed the unusual-looking Bologna stone, that is what he did. Although the stone did not appear to transmutate, it did luminesce for hours afterward.

Modern analysis has now revealed the secret of the stone's afterglow. A team led by scientists from Finland discovered, in 2012, that natural impurities of copper were responsible for the strange glow, taking in energy when exposed to light and then emitting it later. However, they also found that a vital stage was the calcination or heating of the stone, which reduced $BaSO_4$, barium sulfate, to BaS, barium sulfide – so the substance had been transformed after all.

▼ *Glowing Bologna stones.*

Fast Facts

BARIUM [Ba]
ATOMIC NUMBER 56

◄ Witherite is a barium mineral named after William Withering, who also investigated the medical role of the foxglove flower.

Character: Highly poisonous, barium is a fairly soft, silvery metal that is never found in a pure form in nature because it is so reactive. However, as barium sulfate it is widely used as a contrast agent to help take clear X-rays. In this combination, it does not dissolve in water, so the body simply processes it out.

Discovery: Dense, silvery-white pebbles caught the eye of Vicenzo Cascariolo in Bologna, Italy, in 1603. He heated them to a red glow in a heap of charcoal, and found they continued to shine in the dark after exposure to light. Carl Wilhelm Scheele later studied the intriguing stones and realized they were a compound of an unknown element, which turned out to be barium.

► The mineral barite yields barium for commercial purposes.

Name: Barium comes from the Greek word *barys*, meaning heavy. Minerals of barium, such as witherite, discovered in lead mines in northern England by geologist William Withering in 1784, are dense and surprisingly heavy.

World sources: China, India and Morocco have reserves of barite, the main commercial source of barium. Barite is barium sulfate, used in the drilling industry to increase the density of drill fluids.

► Barite crystals mixed with sand form a flower-shaped desert rose.

Fast Facts

RADIUM [Ra]
ATOMIC NUMBER 88

Character: Radium is soft and silvery, and intensely radioactive, glowing with a faint blue light in the dark.

Discovery: Marie and Pierre Curie discovered the element in 1898 in pitchblende, a uranium ore mined in North Bohemia, now in the Czech Republic. Even with the uranium extracted, the remainder was still highly radioactive.

Name: The name radium comes from the Latin word *radius* meaning ray, in reference to the rays of energy the element emitted.

World sources: Radium is found in all rocks containing uranium, and those from the Democratic Republic of Congo, and Canada, are the richest sources. Radium can also be extracted from spent nuclear fuel rods, but annual production is less than 100g (3.5oz).

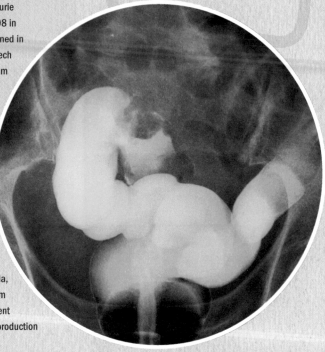

▲ *Barium sulfate in the gastrointestinal tract helps identify a case of colon cancer in this X-ray.*

RADIOACTIVITY REVEALED IN RADIUM AND POLONIUM

When French scientist Henri Becquerel discovered that uranium salts could turn a photographic plate black, he was seeing the first evidence of radioactivity. The uranium affected the heavily shrouded plate via what came to be called Becquerel rays.

It was 1896. Becquerel had been investigating uranium in order to see if it might emit X-rays, which had been announced a few months earlier to great fanfare. He found that the new rays could make air conduct electricity, and had some penetrating effect on materials – so he thought they were a kind of X-ray themselves.

The new rays caught the attention of Marie Curie. She made them the focus of her doctoral research, and took intricate measurements of the ionizing effect of as many uranium compounds as she could, using an electrometer made by

▲ *Pierre and Marie Curie.*

◄ *The Curies used an electrometer invented by Pierre to measure tiny electrical changes caused by radioactive samples.*

her husband Pierre. The conclusion was that the amount of "activity", as Marie called it, depended on the amount of uranium rather than its physical or chemical state.

By testing other elements, she found only one other, thorium, gave off similar ionizing rays to uranium. In 1898, she named this phenomenon radioactivity – a property that Marie correctly believed was intrinsic to the atom itself.

Now she started to look for evidence of undiscovered radioactive elements – and returned to uranium ores, which she had previously observed seemed more radioactive than was warranted by their content of uranium or thorium alone. Pierre joined the search, and they arranged to process tonnes of uranium ore, separating out the most radioactive of the substances it contained.

They were eventually able to announce not one, but two new elements, which they called polonium and radium. For a final flourish, Marie spent four years processing enough of the ore to isolate a tenth of a gram of radium, from which she could determine its atomic weight and confirm its status as a new element.

In 1903, the Curies and Becquerel jointly received the Nobel Prize for Physics for their work on radioactivity.

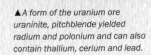

▲ A form of the uranium ore uraninite, pitchblende yielded radium and polonium and can also contain thallium, cerium and lead.

◀ Henri Becquerel saw the evidence of a new kind of radiation in 1896, when crystals of radioactive uranium salt made this photographic plate turn dark.

PERIODIC PIONEERS
MARIE CURIE

Maria Skłodowska – later Marie Curie – was born in Warsaw in 1867 at a time when the city was part of the Russian Empire, and women could not access higher education.

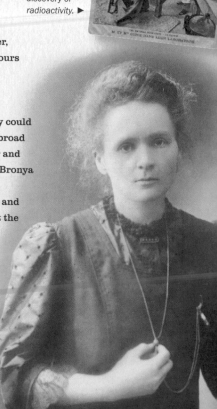

Marie and Pierre Curie received the Nobel Prize for Physics in 1903 for their discovery of radioactivity. ▶

However, she forged the most glittering career, carried out the earliest work on treating tumours with radiation, and is the only person ever to receive Nobel recognition in both physics and chemistry.

Although she loved to learn, Marie's family could not afford to send her, or her sister Bronya, abroad to university. Marie agreed to work as a tutor and governess, reading textbooks at night, while Bronya left to study in Paris.

In 1891, Marie joined her sister in France, and continued her studies in science and maths at the prestigious Sorbonne University. She was an impoverished student, but in 1893 finished top in her masters class in physics. The following year she completed a masters course in chemistry, too – and met Pierre Curie, who she married in 1895.

Marie Curie won Nobel Prizes in both physics and chemistry – a feat that is so far unmatched. ▶

The couple encouraged each other in their scientific work. Marie later said that they were "...so closely united by our affection and our common work that we passed nearly all of our time together."

After labouring over the processing of tonnes of ore, Marie and Pierre announced two radioactive elements to the world in 1898. Marie received her Ph.D in 1903 as well as the Nobel Prize for Physics, shared with Pierre and Henri Becquerel in recognition of their discovery of radioactivity.

Pierre died in 1906 in a road accident. Marie accepted the offer from the Sorbonne to take up his place as professor of Physics. In 1911, Marie received a second Nobel Prize for Chemistry for the discoveries of radium and polonium back in 1898.

▲ Marie Curie and her daughter Irène Joliot-Curie are the only mother and daughter both to have won Nobel Prizes. Irène won the 1935 Nobel Prize for Chemistry.

During the First World War, Marie became head of the Red Cross Radiological Service. She and her elder daughter Irène developed a mobile X-ray machine to diagnose wounded soldiers. Irène was only 17 but was following her mother into a scientific career, and went on to win a Nobel Prize for Chemistry herself.

By 1922, Marie Curie was running the Curie Institute, where she successfully pioneered radium as a treatment for cancer. However, her earlier exposure to radiation had taken its toll. She died at the age of 66 in 1934.

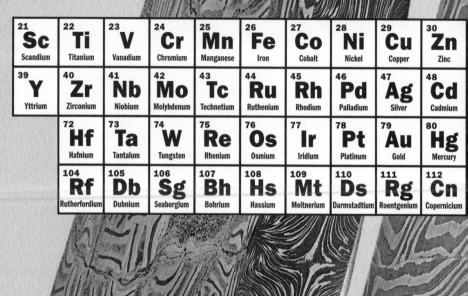

21 **Sc** Scandium	22 **Ti** Titanium	23 **V** Vanadium	24 **Cr** Chromium	25 **Mn** Manganese	26 **Fe** Iron	27 **Co** Cobalt	28 **Ni** Nickel	29 **Cu** Copper	30 **Zn** Zinc
39 **Y** Yttrium	40 **Zr** Zirconium	41 **Nb** Niobium	42 **Mo** Molybdenum	43 **Tc** Technetium	44 **Ru** Ruthenium	45 **Rh** Rhodium	46 **Pd** Palladium	47 **Ag** Silver	48 **Cd** Cadmium
72 **Hf** Hafnium	73 **Ta** Tantalum	74 **W** Tungsten	75 **Re** Rhenium	76 **Os** Osmium	77 **Ir** Iridium	78 **Pt** Platinum	79 **Au** Gold	80 **Hg** Mercury
104 **Rf** Rutherfordium	105 **Db** Dubnium	106 **Sg** Seaborgium	107 **Bh** Bohrium	108 **Hs** Hassium	109 **Mt** Meitnerium	110 **Ds** Darmstadtium	111 **Rg** Roentgenium	112 **Cn** Copernicium

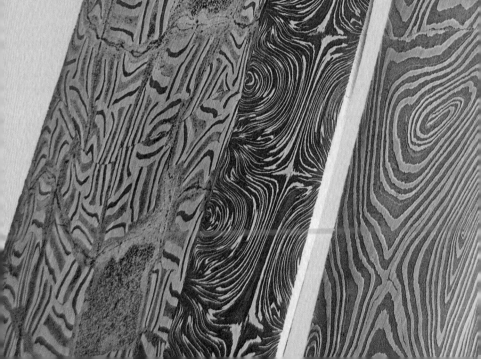

CHAPTER 5

TRANSITION METALS

WHICH TRANSITION METALS CREATED THE SECRET OF DAMASCUS STEEL SWORDS?

HOW DOES TITANIUM KEEP US ALL IN PLASTICS?

WHY WAS AN EARTHQUAKE SO SIGNIFICANT TO THE CAREER OF INGE LEHMANN?

METAL MYSTERIES

Among the 118 elements in the periodic table, a large majority are metals. Yet, if you look up the meaning of the word "metal", it is not a very clear-cut definition.

Metals are *generally* malleable, textbooks say. You can draw them out into a wire. They are *typically* opaque, dense, and hard at room temperature – although not mercury, of course, which does not freeze until it is cooled to -38.83°C (-37.89°F).

Metals, we read, also have a metallic lustre. So, metals...appear metallic?

▲ *Aluminium, brass, steel and copper: the transition metals make up some of the most useful substances we know.*

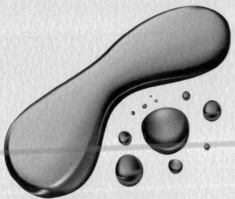

▲ *Mercury is the only metal liquid at room temperature.*

We can definitely all agree that metals conduct electricity. This is because they have a particular kind of bonding in which their outer electrons are free to flow. Their atoms are held together with strong bonds, which is why they have relatively high melting and boiling points (except mercury, again). (And gallium, which will melt in your hand.)

We know a metal when we see one. But to understand the group called the transition metals, we have to look deeper.

1 H																	2 He
3 Li	4 Be											5 B	6 C	7 N	8 O	9 F	10 Ne
11 Na	12 Mg											13 Al	14 Si	15 P	16 S	17 Cl	18 Ar
19 K	20 Ca	21 Sc	22 Ti	23 V	24 Cr	25 Mn	26 Fe	27 Co	28 Ni	29 Cu	30 Zn	31 Ga	32 Ge	33 As	34 Se	35 Br	36 Kr
37 Rb	38 Sr	39 Y	40 Zr	41 Nb	42 Mo	43 Tc	44 Ru	45 Rh	46 Pd	47 Ag	48 Cd	49 In	50 Sn	51 Sb	52 Te	53 I	54 Xe
55 Cs	56 Ba		72 Hf	73 Ta	74 W	75 Re	76 Os	77 Ir	78 Pt	79 Au	80 Hg	81 Tl	82 Pb	83 Bi	84 Po	85 At	86 Rn
87 Fr	88 Ra		104 Rf	105 Db	106 Sg	107 Bh	108 Hs	109 Mt	110 Ds	111 Rg	112 Cn	113 Nh	114 Fl	115 Mc	116 Lv	117 Ts	118 Og

▼The d-block transition metals, highlighted within the periodic table.

s-block   **d-block**   **p-block**

57 La	58 Ce	59 Pr	60 Nd	61 Pm	62 Sm	63 Eu	64 Gd	65 Tb	66 Dy	67 Ho	68 Er	69 Tm	70 Yb	71 Lu
89 Ac	90 Th	91 Pa	92 U	93 Np	94 Pu	95 Am	96 Cm	97 Bk	98 Cf	99 Es	100 Fm	101 Md	102 No	103 Lr

f-block

THE D-BLOCK ELEMENTS

The transition metals form the largest set of elements in the periodic table. They make up four rows, or periods, of elements from columns 3 to 12. They are also known as d-block elements, and this starts to unlock their true nature.

Quantum theory says that electron shells are split into orbitals, starting with the s orbital, then p, and then d, and so on. This helps explain the reactivity and other behaviour of sets of elements. We can see the periodic table in terms of these orbitals, too.

The s-block contains the elements whose outermost electrons are in s orbitals, and have one or two electrons in the outer shell. These are the alkali metals and alkaline earth metals.

The p-block is on the right side of the periodic table, and incorporates elements in the rightmost six columns. Their outer electrons are in the p orbitals, which can hold a maximum of six electrons, leading to the six vertical groups as the orbitals fill.

The d-block, or transition metals, have their outermost electrons in a d orbital. These orbitals can hold up to five pairs of electrons, so there are ten columns.

▲ The transition metals are useful catalysts in many industrial processes.

Transition metals have lots of options when it comes to bonding. They can take on many possible oxidation states by losing electrons and becoming ions, and by acquiring incompletely filled d orbitals. Iron, for example, can have many different oxidation states, although it tends to form compounds in the +2 and +3 states.

CATALYSTS FOR CHANGE

The transition metals' flexibility in taking on oxidation states means they are extremely useful, making various reactions easier or cheaper by catalyzing them without being changed themselves.

Sulfuric acid is a key industrial ingredient for manufacturing fertilizers and pigments, and in metal processing. In the contact process for making sulfuric acid, a vanadium catalyst increases the rate of reaction. Platinum was used in earlier years, but tended to lose its potency if it reacted with impurities.

Ammonia production is also required on a colossal scale for the agriculture, dye and plastics industries. The Haber process involves reacting hydrogen and nitrogen at high temperature and pressure – in the presence of an iron catalyst to speed things along.

Polythene, the most common plastic, can be made at lower temperatures and pressures because of a titanium-based catalyst, a discovery that won its German inventor, Karl Ziegler, the Nobel Prize for Chemistry in 1963.

▲ Metals conduct electricity because their outer electrons are free to flow.

▲ Polythene production relies on a transition metal catalyst, titanium.

◄ Like many transition metals, vanadium can exist in several different oxidation states that produce solutions of different colours.

115

Fast Facts

SCANDIUM `Sc`

ATOMIC NUMBER 21

Character: Scandium is a slightly shiny silver-coloured metal. It is combustible, reacts with water and is lightweight.

Discovery: Dmitri Mendeleev predicted that there ought to be an element with an atomic weight between calcium and titanium, and left a gap for it. Sure enough, Lars Fredrik Nilson discovered Mendeleev's missing element in 1879. He extracted it from a mineral called euxenite, a Greek name meaning "welcoming to strangers", because it is a source of several other elements too.

Name: Nilson was Swedish, and working at the University of Uppsala where he encountered this new element, so he named it scandium – the Latin name for Scandinavia. He was a former student of Jacob Berzelius, who had found four new elements himself, and published an influential table of atomic weights in 1818.

World sources: Because of the complexity of obtaining scandium from ore, it was not until the 1960s that chemists managed to produce the first one pound (nearly half a kilo) block of pure scandium. The element exists in over 800 different minerals distributed throughout the world, but usually only in tiny quantities.

▲ Scandium.

▲ Scandium and titanium make lightweight bicycle frames.

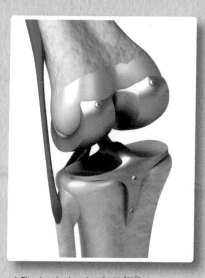

Fast Facts
TITANIUM Ti
ATOMIC NUMBER 22

▲ *Titanium is strong and non-toxic, ideal for making artificial joints.*

▼ *It is also added as titanium dioxide to whiten cosmetics and toothpaste.*

Character: Titanium has a silvery complexion and many charms. It is very light and yet exceptionally strong, it never corrodes, and it is not magnetic. Its down-to-earth uses in toothpaste, sunscreen and artificial joints contrast with high-flying specialist applications in the aerospace industry.

Discovery: William Gregor discovered titanium in a black sand he called menachanite after the Menchan Valley in Cornwall, southwest England, in 1791. Soon after, it was also found by Martin Klaproth, a leading German chemist.

Name: Martin Klaproth named titanium after the Titans and Titanesses, mythical Greek gods who were children of the gods of the earth and sky.

World sources: A New Zealander, metallurgist Matthew Hunter, separated pure titanium in 1910 while working in America at the General Electric Company. The process of isolating the pure metal remains difficult, even for modern scientists, and this helps explain the element's high price, even though it is fairly abundant. Canada, South Africa and Australia are the world's largest producers.

◀ *Titanium.*

Fast Facts
VANADIUM [V]
ATOMIC NUMBER 23

Character: Vanadium is a silvery-white metal predominantly used to make steel stronger and shatterproof. Used in jet engines and nuclear reactors, its strength and lightness also come into play in military helmets and body armour.

Discovery: Strangely, vanadium had to be discovered at least twice before anyone in the scientific community would accept its existence as a new element. Spanish-Mexican mineralogist Andrés Manuel del Rio was arguably the first to find it. He sent a sample from his lab in Mexico City to Paris for analysis in 1801 – but it was dismissed as a chromium mineral. Another 30 years passed before Swede Nils Gabriel Sefström extracted a new element from iron and called it vanadium.

Then in 1869, British scientist Henry Roscoe obtained the first specimen of almost pure vanadium in an extraction process he perfected in Manchester. He was also able to establish that the previous discoveries in Mexico and Sweden had actually been vanadium compounds.

Name: Vanadium takes its name from Vanadis, the goddess of beauty in Norse mythology.

World sources: The world's largest producers of vanadium from mineral extraction are Russia and South Africa.

▲ Vanadium is strong and light enough to make body armour.

▲ Steel tools can be toughened by adding vanadium.

Fast Facts

CHROMIUM [Cr]

ATOMIC NUMBER 24

Character: Hard, brittle chromium makes a smooth silvery plating, and one of its key uses is as an ingredient in stainless steel. It is an essential trace element in our diet, but contact with chromium salts causes hideous skin ulcers.

Discovery: A Siberian miner never realized that he had struck gold – figuratively speaking – in the 1760s. His pickaxe dislodged a deep red nugget that German geologist Johann Gottlieb Lehmann named Siberian red lead. In 1798, French chemist Nicolas-Louis Vauquelin investigated red lead, and managed to separate a new element, chromium.

▲ *Chromium.*

Name: Chromium causes an array of natural vibrant colours, including the green in emeralds and red in rubies. Struck by the palette chromium could offer, Vauquelin named the new element after the Greek word for colour: *chroma*.

World sources: The two biggest producers of chromium are South Africa and Kazakhstan. It is also found in India, Brazil, Finland, Albania, Turkey and the USA. The frozen north – particularly Greenland, as well as Canada, Russia and Finland – has been identified by geologists as a likely source of unexplored reserves of chromium.

▲ *The colour of a ruby is created by the element chromium.*

▲ 17,000 years old, or more, the Lascaux cave paintings in France depict animals and other figures using manganese pigments.

Fast Facts

MANGANESE Mn
ATOMIC NUMBER 25

Character: Manganese is brittle, hard and has a darkish silvery colour. Its main application is in steel manufacturing, in which it converts iron sulfide to manganese sulfide, allowing the steel to be rolled and forged.

Discovery: In 1774, Swedish chemist Carl Wilhelm Scheele presented evidence that manganese existed as a separate element to magnesium, which often turned up in the same minerals. His colleague Johan Gahn eventually purified manganese and secured its elementary status.

Name: Manganese may have come from the name for the black ore in which it was found – magnesia nigra, from a place called Magnesia in modern Greece. On the other hand, it may refer to the Latin for magnet, *magnes*. Manganese is non-magnetic, but a salt, manganese sulfate, does demonstrate magnetic properties.

World sources: Only four elements are found in more common abundance in the Earth's crust than manganese. It has also been discovered on the bottom of the ocean. Gabon, as well as Ukraine, Australia, South Africa and China, is currently among the world's top producers.

Pure manganese. ▲

Fast Facts

IRON [Fe]
ATOMIC NUMBER 26

Character: Iron is a hard yet workable and ductile metal that rusts quickly. Ninety per cent of all refined metal in the world is iron, and we depend on it as the principal component of manufactured steel.

Discovery: Humans have been using iron for at least 5,500 years, ever since the Egyptians stumbled across useful meteorite deposits of the element. French scientist René Antoine Ferchault de Réaumur began our latest Iron Age when he investigated the differences between iron and steel – principally their carbon content – and kick-started the Industrial Revolution.

Name: The word iron has evolved slightly from the Old English word *iren*. It appears as Fe in the periodic table, derived from *ferrum*, the Latin word for iron. The adjective ferrous was coined in the 19th century to describe any metal that contains iron.

World sources: Iron is actually the Earth's most abundant element by mass, but most is locked up in the core. However, it is found as a more accessible natural deposit across a vast array of nations, and China, Australia and Brazil are leading producers.

▼ *County Kerry in Ireland is the site of this fort from the Iron Age, when people began using iron to make tools and weapons.*

▲ *The world's first iron bridge crossed the UK's River Severn in 1779 and brought in an era of cast-iron bridges and buildings.*

WOMAN OF IRON

When Jules Verne published his science fiction novel _Journey to the Centre of the Earth_ in 1864, it is fair to say that humanity did not have much of a clue as to what lay at our planet's core. That revelation came 72 years later, when a groundbreaking Danish seismologist made some key calculations.

Inge Lehmann had studied maths, and developed an interest in earthquakes. She knew that seismological data held the key to discerning the structure of the ground beneath our feet. In 1925, she began working to set up seismological observatories in Denmark and Greenland.

As earthquake vibrations rumble through the Earth, some move as primary, or pressure, waves (P-waves), while others are secondary, or shear, waves (S-waves). The pattern of waves arriving at monitoring stations on the planet's crust reveals information about their routes and the materials they have passed through. Also, crucially, only P-waves can travel through both solids and liquids.

The prevailing idea was that the core of the Earth was entirely molten iron. However, Lehmann spotted a problem with the data, particularly as she studied

▲ _After the Murchison earthquake in New Zealand in 1929, a team tries to free this car from a crack in the ground._

▲ *Inge Lehmann.*

seismic arrivals following the Murchison earthquake that struck New Zealand in 1929. The P-waves were moving through the Earth in ways that theory could not explain.

Lehmann created mathematical models of an Earth with a solid inner core and a liquid outer core, and wrote a crucial scientific paper. Published in 1936, it overturned the belief that our planet was so hot below its solid crust that the core must be completely molten – and proved what Jules Verne would really find on a journey to the centre of the Earth.

SECRETS OF DAMASCUS STEEL

According to legend, Damascus steel swords were so sharp they could cut a hair if it simply fell across the blade. During the 17th century, the weapons were widely prized, but nobody could persuade a steelmaker to reveal their secrets.

It is only in recent years that the magical metal has given anything away. The steel, called wootz in English, originated in India. It has unique banding patterns formed by iron alloys mixed with carbon nanowires and nanotubes. Trace amounts of tungsten or vanadium are also present – and the fact that production of this super-material ended around 1750 may be because the source of ore that provided the crucial mixture simply ran out.

◀ *Damascus steel blade showing swirling patterns.*

Fast Facts
COBALT [Co]
ATOMIC NUMBER 27

Character: Cobalt is a strong, brittle, silvery-white metal, which can be magnetized like iron.

Discovery: The blue glass stripes in the famous golden mask of the Egyptian pharaoh Tutankhamun, who died in 1323 BCE, owe their colour to cobalt. But it was not until well into the 18th century that anyone suspected the substance responsible might be an individual element. Swedish chemist and metallurgist Georg Brandt studied intriguingly blue-coloured ore from a local copper mine at Västmanland. He managed to isolate cobalt in 1739, becoming the first person to find a new metal since prehistory.

Name: Cobalt comes from the German word for goblin. Frustrated miners would extract cobalt ore, hoping to be able to smelt valuable metals such as nickel, copper or even silver. However, as the ore also contained arsenic, the process often resulted in clouds of troublingly toxic vapour – and no precious metal. Goblins got the blame.

World sources: Sixty per cent of the world's cobalt comes from the Democratic Republic of Congo. Demand is on the rise for use in rechargeable batteries – but conditions tend to be dangerous and poorly rewarded for the workforce of miners in one of the world's historically most exploited countries.

◄ *Cobalt blue, a pigment containing cobalt and aluminium, is a popular colour for porcelain, tiles and glass.*

COBALT AND TRADE ROUTES

From ancient China to Iraq and Egypt, artisans have long recognized the beauty of cobalt blue. But in 2016 researchers made an unexpected find while examining Bronze Age burials far away to the northwest, in Denmark.

In two graves at Ølby, about 40km (24.85 miles) south of Copenhagen, the scientists found jewellery made from both amber and blue glass beads, probably representing a tribute to the Sun. When the researchers used plasma spectrometry to examine the blue beads, they had a surprise: their chemical composition was the same as glass found in Egyptian tombs. The team even concluded that the beads had originated in the same workshops that made the death mask of the boy-king Tutankhamun.

The finding is evidence that Denmark and Egypt traded together over 3,000 years ago, with Nordic amber travelling south and cobalt jewellery travelling north.

▲ The same Egyptian workshops that made Tutankhamun's mask also produced the blue glass beads found in a Bronze Age grave in Denmark.

THE POWER OF MAGNETISM

For 800 years, Olmec civilization flourished in the tropical lowlands we now call Guatemala and Mexico. Its people made carvings, played ball games, raised crops and built waterways. No one is sure how many of the Olmec customs and rituals passed into the cultures that displaced them in about 400 BCE. But it is possible that the Olmec people were the earliest ever to make use of magnetism.

An archaeologist investigating an Olmec site called San Lorenzo in 1966 came across a very modern-looking artefact. It was a flattened, oblong, polished bar of magnetic haematite – an iron-containing ore, iron(III) oxide – with a groove along its length. One researcher speculated it could have been part of a very early compass.

Far-fetched, perhaps. But in the 1970s, stone sculptures with magnetic properties began turning up in further investigations of Olmec sites. The carvings depicted people and animals, and appeared to have been deliberately made to highlight the magnetite ore naturally present in the original basalt boulders. By using modern compasses, researchers found

◄ *Olmec civilization is famous today for colossal sculptures of heads, which archaeologists have unearthed from across the landscape.*

that sculptures of human heads had a concentration of magnetic lines of force that entered above the right ear, and emerged below it. A turtle head sculpture, meanwhile, had a strong north pole in its snout and a south pole at the back.

The Olmec were a Stone Age people without knowledge of metalworking. Yet they seem to have appreciated the invisible power of magnetism. Exactly how, and why, remains a mystery – but in the timeline of history, another thousand years passed before the first records of cultures systematically using magnetic materials for way-finding and sea navigation.

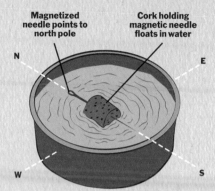

Magnetized needle points to north pole

Cork holding magnetic needle floats in water

Simple Compass

▲ Navigation using magnetic compasses only began a thousand years after the Olmec's magnetic sculptures.

CHINESE COMPASSES

Zhu Yu's book *Pingzhou Table Talks* is the earliest record of a magnetic compass used for maritime navigation. The author refers to a south-pointing needle, which would have been magnetized by rubbing with a lodestone. The book dates from 1111–1117 during the Song Dynasty in China. In fact, earlier Chinese compasses had existed since the 2nd century BCE, but people used them to tell fortunes rather than for travel.

▲ Chinese feng shui practitioners used compasses made using naturally magnetic lodestones as early as the 2nd century BCE.

TRANSITION METALS AND MAGNETISM

To understand what makes things magnetic, we have to delve again into the quantum-level world of electrons. Other than their electrical charge, electrons also have another fundamental property, called spin, which produces a tiny magnetic field. A pair of electrons in a shell will have one spin up, and one spin down, cancelling each other out.

Any element with unpaired electrons, like most of the transition metals, are **paramagnetic** materials. When an external magnetic field acts on them, their unpaired electrons can align with the field, reinforcing it and causing a weak net attraction. Some transition metals go the other way and become **diamagnetic** – weakly repelled by an applied magnetic field – including mercury, gold and copper.

But the most familiar everyday magnets are made from iron, nickel and cobalt, which exhibit **ferromagnetism**. As well as being paramagnetic, these elements' electrons tend to spontaneously

FLYING FROGS

If you have ever wanted to levitate, then diamagnetic forces may hold the key. Diamagnetism is the property all materials have to repel an externally applied magnetic field by creating one of their own. The effect is weak and usually masked by other forces. But in the 1990s, Andre Geim decided to put water's diamagnetic properties to the test. He set an electromagnet to full power and poured water into it. To Geim's surprise, spheres of water started to float, their diamagnetism overcoming the force of gravity. He and colleagues then experimented with a frog, and – in a truly eyebrow-raising moment – found that it, too, levitated above the magnet.

Andre Geim and his collaborator Michael Berry won the Ig Nobel prize for physics in 2000, one of ten annual awards given for achievements that first make people laugh, and then make them think. Since then, his unusually wide-ranging experimental approach has also led to him winning the 2010 Nobel Prize for Physics for his work discovering graphene.

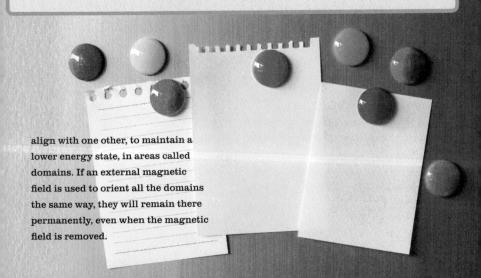

align with one other, to maintain a lower energy state, in areas called domains. If an external magnetic field is used to orient all the domains the same way, they will remain there permanently, even when the magnetic field is removed.

MAGNETITE AND HAEMATITE

◄Could lightning strikes be responsible for making unexpectedly strong natural magnets?

What about the naturally magnetic iron ore, magnetite, which the Olmec seemed to know about in their sculptures? This is an example of a ferrite, a group of substances consisting of iron oxides plus one or more extra metallic elements. They gave their name to ferrimagnetism, only discovered in 1948: a type of weak permanent magnetism in which neighbouring pairs of electrons point in different directions in their domains, but do not entirely cancel each other out.

Magnetite is the material from which lodestones were made, the earliest natural magnets used by navigators. However, ferrimagnetism does not explain the strength of a lodestone's field. Some scientists have suggested that lightning strikes may be responsible for the powerful magnetic fields required to align domains in the substance. But crucially, tiny regions of other oxides, or metals such as titanium, aluminium or magnesium within the magnetite, stop the domains from returning to a more disordered state when the external field disappears.

And finally, how can we explain the carefully polished Olmec bar, made of haematite (see page 126)? Although haematite is iron(III) oxide, it is not usually noticeably magnetic. Why would this ancient sample have had a stronger field? Detailed analysis showed it had titanium-containing layers within its structure, and its permanent magnetism may have arisen in a similar way to that of lodestones. More, we cannot currently say.

Fast Facts
NICKEL [Ni]
ATOMIC NUMBER 28

Character: Nickel is a malleable, bright silver metal often used in decorative plating – although in jewellery it can result in skin reactions.

Discovery: A significant proportion of nickel arrived on Earth from outer space, in the form of a meteorite that fell on Canada. Some is also present deep underground. It was not until 1775 that the mineral was purified by Swedish chemist and mineralogist Torbern Bergman.

Name: The word nickel is a corruption of the German term *kupfernickel*, meaning "copper devil". Hardworking 18th-century German miners sometimes spent hours working a rich seam of red ore, only to discover it was nickel instead.

World sources: Nickel is commercially extracted in over 20 countries. Canada, Russia, Colombia, Australia, Indonesia and the French Pacific territory New Caledonia, are among the biggest annual producers of the element.

▲ *NASA built the Space Shuttle Discovery's main engine from 65 per cent nickel-based alloys because of their ability to resist corrosion.*

Margarine manufacturers use a nickel catalyst to hydrogenate liquid vegetable oil, turning it into solid spread. ▶

NITINOL MEMORY METAL

In 1961, when US scientist William Buehler was researching metallic alloys for new Navy missile nose cones, his investigations hit an important target – but not the one he was aiming for. In pursuit of a fatigue-resistant substance, he created a mixture of nickel and titanium and called it nitinol. He took strips of nitinol to demonstrate at a meeting, folded multiple times to show its ability to unfold again without damage. One of his bosses unexpectedly took a pipe lighter, heated the compressed nitinol strip – and before everyone's eyes, it stretched out again to its original shape.

Today, shape-memory alloys like nitinol are useful for medical stents, self-adjusting dental wires, spectacle frames, and intra-uterine contraceptive devices. The way memory metals work is by changing phase between two solid states. At lower temperatures, the crystal structure is called the martensite phase, a less symmetric and less dense phase, in which the nitinol can be bent into different shapes. When the alloy is heated, it transforms from martensite to the austenite phase, a symmetric crystal structure, and returns to the shape it had before it was deformed. The transition temperatures can be fine-tuned by adjusting the nickel-to-titanium ratio.

▲ *Glasses frames made from shape-memory alloys like nitinol remember their shape after being bent.*

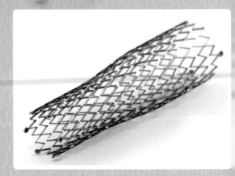

Self-expanding stents made from nitinol are vital in keyhole surgery for repairing arteries. ▶

▶ *Once the fastest person to sail solo around the world, Ellen MacArthur is now a circular economy pioneer.*

CIRCULAR ECONOMY

British sailor Ellen MacArthur broke the world record for the fastest solo circumnavigation of the globe in 2005, at the age of only 28. Today, however, she is more interested in completing a different kind of circuit. She now works with businesses, politicians and researchers to advocate for a circular economy. One important aspect of this new way of thinking is that raw materials such as copper, gold, platinum, lithium and aluminium are recovered and recycled.

▼ *In the circular economy, engineers design gadgets so that their raw materials can be harvested instead of being thrown away.*

Current manufacturing often involves taking raw materials, manufacturing goods and then throwing them away later. Under a circular economy, designers and engineers plan the whole life of a product, including materials, energy and its end-of-life reuse. Apple is one company working on its circular economy ambitions. In 2016, the company announced a new robot called Liam, designed to disassemble iPhones to harvest useful metals hidden away inside, all of which can be recycled in new devices. The project is still in its early stages.

Fast Facts

COPPER $\boxed{Cu}$
ATOMIC NUMBER 29

Character: Copper is an attractive reddish-brown colour – but only for a short while. It quickly oxidizes to black copper oxide and then develops a distinctive green patina known as verdigris. Widely used for its electrical conductivity, copper also conveys heat very effectively.

Discovery: Copper exists in its pure form in nature, and so is probably the first metal that people extracted and shaped into tools, weapons and jewellery. Later, the breakthrough discovery that adding tin to copper created a stronger, harder metal brought in the Bronze Age around 3500 BCE.

Name: The word copper derives from the name the Romans gave to Cyprus. Such was copper's historical association with the Mediterranean island that it was named *Cyprium aes*, meaning "a metal from Cyprus".

World sources: In 1860, Swansea, on the south coast of Wales, was the world's leading copper provider. Nowadays copper is produced in over 50 countries, with Chile being one of the largest providers. USA, China, Peru, Japan and Mexico also dig and smelt huge quantities.

◄ Bright green malachite is a copper mineral.

▲ Horseshoe crabs have blue blood because it contains copper.

◄ Sheets of copper on the Statue of Liberty have developed the now-familiar green patina.

Fast Facts

ZINC [Zn]
ATOMIC NUMBER 30

Character: A whitish, silvery metal with a subtle blue tinge, zinc is the substance with which iron is plated to prevent rusting in a process called galvanization. Rolled zinc sheeting was the primary material used for roofing in Paris in the 1860s, inspiring painters including Cézanne and Van Gogh.

Discovery: Zinc has a long history. In the 1st century CE, Pliny the Elder wrote about a soothing ointment that contained zinc oxide. A large-scale zinc smelting operation functioned in Zawar, Rajasthan, for around 400 years between 1100 and 1500. However, while a Flemish metallurgist, P Moras de Respour, reported the extraction of metallic zinc in 1668, the person who recognized zinc as a new element was the German chemist Andreas Marggraf in 1746.

Name: The word zinc is probably a straight import into the English language from the identical term in German vocabulary. However, some scholars believe the German word zinc originated from the Persian language as a derivative of *sing*, meaning stone.

World sources: China, Australia and Peru are currently the world's largest producers of zinc, with Canada, USA, Iran, India, Mexico and Ireland also making contributions.

▲ Zinc roofs of Paris.

Fast Facts
YTTRIUM ⟨Y⟩

Character: Yttrium is a soft, silvery, shiny metal that does not exist in pure form in nature but is present in several minerals.

Discovery: A surprisingly heavy black rock was the first sign of the new element yttrium. A Swedish army officer called Karl Arrhenius stumbled across it in 1787, and called it ytterbite after the nearby village of Ytterby. He passed the mineral to a Finnish chemist called Johan Gadolin to investigate, thinking it contained tungsten.

Name: Johan Gadolin identified a new oxide in the weighty rock sample. The oxide came to be called yttria – but later, this yielded not only yttrium, purified by Friedrich Wöhler in 1828, but also further elements ytterbium, terbium and erbium. *Ytterby* translates as "the outer village" and is found a ferry ride away from present-day central Stockholm.

Gadolin's fame is preserved in the name of the original heavy coal-like mineral, now called gadolinite.

World sources: China, Malaysia, Russia and Madagascar.

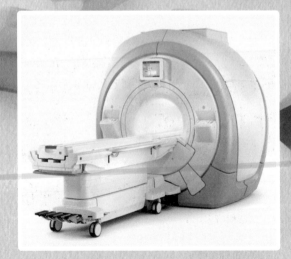

▼ *Yttrium aluminium garnets make convincing synthetic diamonds and coloured gems.*

◄ *High-temperature super-conductors made from an yttrium compound offer the hope of cheaper magnets for MRI machines.*

LIFE-SAVING SUPERCONDUCTORS

Electrical current can flow through a metal because the electrons are free to move. But it is not very efficient: energy is lost to resistance through collisions and vibrations. At a low enough temperature, some materials enter a superconducting state in which a current can flow indefinitely with zero resistance. Dutch physicist Heike Kamerlingh Onnes of Leiden University discovered the phenomenon in mercury, back in 1911. He cooled the sample using helium that he had only recently worked out how to liquefy – the first person to do so.

As experiments continued, scientists discovered that niobium metal, in combination with tin, could be made into superconducting wires for powerful electromagnets. Niobium-titanium worked even better, and became the main material used in medical magnetic resonance imaging (MRI) machines, cooled with liquid helium down to -269°C (-452.2°F).

However, helium is an expensive coolant. In 1986, scientists based in Zurich, Georg Bednorz and Karl Müller, discovered materials that could superconduct at higher temperatures, and won the 1987 Nobel Prize for Physics for their work. Soon afterward, another breakthrough came along in the form of yttrium barium copper oxide (YBCO) compounds, which

were the first materials found to superconduct slightly above the boiling point of readily available liquid nitrogen (-196°C/-320.8°F).

▲ *The Large Hadron Collider project at CERN requires superconducting cables, here constructed from niobium and tin.*

Fast Facts
ZIRCONIUM [Zr]
ATOMIC NUMBER 40

Character: Alphabetically last in the periodic table, zirconium is a hugely important element for 21st-century technology. Pure zirconium is a tough, hard, light-grey metal that is usefully resistant to corrosion and can withstand conditions inside furnaces and nuclear reactors.

Discovery: Zirconium's best-known guise is as cubic zirconia, a gemstone easily confused with diamonds. German chemist Martin Klaproth devised a test in 1789 to distinguish the two. He did not manage to separate elemental zirconium, but an oxide. Jacob Berzelius had more success in 1824, and finally Dutch scientists Jan Hendrik de Boer and Anton Eduard van Arkel produced pure zirconium in 1925.

Name: The element's name is a corruption of the Arabic word *zargun* that means gold-coloured, in reference to its golden crystals.

World sources: Australia, South Africa and China produce the majority of the world's zirconium ore, called zircon.

▼ *Zirconium alloys clad the fuel rods in nuclear reactors because they are highly resistant to heat and to chemical corrosion.*

▲ *Hard, colourless cubic zirconia is the cubic crystalline form of zirconium dioxide, and in the past was hard to distinguish from a diamond.*

NIOBIUM Nb
ATOMIC NUMBER 41

Character: Niobium is a silvery metallic element that is resistant to corrosion, and has useful superconductive properties.

Discovery: While arranging minerals for display in the British Museum in London, chemist Charles Hatchett noticed what he later recorded as "a small specimen of a dark-coloured heavy substance" in the museum's collection. Hatchett owned a chemical works in west London, and experimented on a sample of the rock, reporting in 1801 that he had partially extracted a new element he had named columbine.

Not all chemists were convinced about Hatchett's discovery, especially as an incredibly similar element, tantalum, was identified the following year. But in 1844 German chemist Heinrich Rose proved that columbine was indeed a new element.

Name: Charles Hatchett called the new element columbine in recognition of the fact that the museum sample had originated in America – known sometimes historically as Columbia after Christopher Columbus. However, Heinrich Rose preferred niobium, to recall Niobe, the daughter of King Tantalus in Greek mythology, and in reference to the element's similarity to tantalum metal.

After a period of standoff, German chemists agreed to call some other elements by their international rather than German names and gained the right to rename columbine as niobium.

World sources: Brazil mines the most niobium by a long way, but Canada and Australia also produce the metal.

Niobium is a good choice for jewellery because it is biocompatible. ▶

Fast Facts
MOLYBDENUM `Mo`
ATOMIC NUMBER 42

Character: A strong, hard metal, bright silver in colour, molybdenum is often used as a strengthening alloy.

Discovery: For centuries, an unremarkable black mineral was dismissed as lead ore. In 1778, when Swedish chemist Carl Scheele investigated it, he found it contained a mystery metal instead. Scheele handed the mineral to fellow Swede, Peter Jacob Hjelm, who managed to isolate a new element.

Name: Molybdenum originates from the Greek word *molybdos* meaning lead, in recognition of its long-lasting case of mistaken identity.

World sources: China, USA and Chile.

▲ *Once a mystery metal, molybdenum was identified by Peter Hjelm and announced in 1781.*

Fast Facts
TECHNETIUM `Tc`
ATOMIC NUMBER 43

Character: Technetium was the world's first artificially produced element. However, we now know that minute traces of technetium do occur on Earth.

Discovery: An element called ekamanganese was one of the missing substances for which Dmitri Mendeleev left a space in the periodic table in 1869. When other absentees, including scandium, gallium and germanium, slotted into their various gaps, it made the lack of number 43 even more conspicuous. Finally, in 1937, mineralogist Carlo Perrier and chemist Emilio Segrè tracked down the elusive element in the form of a pair of radioactive isotopes. They turned up in radioactive components from an experiment done by the inventor of the cyclotron particle accelerator, Ernest Lawrence.

Name: Technetium's name arises from the word *tekhnetos*, which means "artificial" in Greek.

World sources: Given the right kit, technetium can be created anywhere.

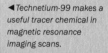
◄ *Technetium-99 makes a useful tracer chemical in magnetic resonance imaging scans.*

▲ Jędrzej Śniadecki discovered ruthenium in 1808, although he could not replicate his findings. He also created the modern chemistry terminology used in Polish.

▼ Lighter, stronger wind turbine blades could be made with resins and ruthenium catalysts.

Fast Facts
RUTHENIUM [Ru]
ATOMIC NUMBER 44

Character: Ruthenium is a hard, silvery metal that is almost untarnishable.

Discovery: Polish chemist Jędrzej Śniadecki was first to find ruthenium in 1808, among South American platinum ore samples. He named the element vestium after an asteroid, but later retracted his discovery when it could not be replicated. Then in 1825 Gottfried Osann, a German scientist based in Estonia, conducted an experiment on platinum ore mined from the Ural Mountains and discovered the same element as Śniadecki, naming it ruthenium. Pure ruthenium was extracted in 1840 by Karl Karlovich Klaus, working at the University of Kazan in Tatarstan.

Name: The name ruthenium derives from *Ruthenia*, the Latin word for Russia.

World sources: South Africa, Russia, USA and Zimbabwe.

Fast Facts
RHODIUM [Rh]
ATOMIC NUMBER 45

Character: Rhodium is a hard, bright silvery-white metal that is strongly resistant to corrosion. It is sometimes credited as the world's most expensive element, and is the rarest of all non-radioactive metals.

Discovery: British metallurgist William Hyde Wollaston discovered rhodium. The Norfolk-born scientist found it in a sample of Colombian platinum which he exposed to acid and created a red solution. From this he made pinkish-red crystals that turned out to be rhodium chloride salts, which he reduced to make metallic rhodium. He recognized it as a new element and announced it as such in 1803.

Name: Rhodium takes its name from the Greek word *rhodon* meaning rose, because of its rose-red crystals.

World sources: Only 30 tonnes of rhodium is produced each year, from deposits in South Africa, Russia, Canada, USA and Zimbabwe. It is extracted as a by-product of copper and nickel refining and is useful in catalytic converters in cars. Japanese scientist Ryoji Noyori won the Nobel Prize for Chemistry in 2001 for his development of catalysts using rhodium, useful in drug production and also in synthesizing menthol flavouring.

▲Rhodium is an active component in car catalytic converters.

野依 良治

▲Japanese chemist Ryoji Noyori developed rhodium catalysts.

▼One use of rhodium catalysts is in synthesizing menthol flavouring, demand for which outstrips supply from natural sources such as peppermint plants.

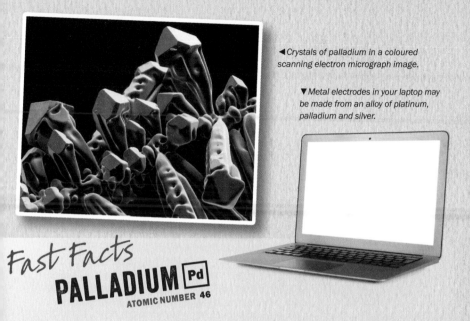

◀ Crystals of palladium in a coloured scanning electron micrograph image.

▼ Metal electrodes in your laptop may be made from an alloy of platinum, palladium and silver.

Fast Facts
PALLADIUM [Pd]
ATOMIC NUMBER 46

Character: Malleable and silvery-white, palladium is a metal that resists tarnishing well.

Discovery: Palladium is the second element discovered by William Hyde Wollaston. Along with rhodium, palladium turned up in a sample of platinum ore. Wollaston was working with another English chemist called Smithson Tennant to investigate the ore, and between them they found four of the six platinum group metals. Tennant's contributions were osmium and iridium.

Name: Wollaston was about to name his new element ceresium after an asteroid called Ceres appeared in 1801 – but then a newer asteroid named Pallas turned up, and he decided to use this instead. Pallas recalls the Greek goddess of wisdom.

World sources: Only about 200 tonnes of palladium is produced globally each year, and a quarter of that figure comes from recycling. Its uses are in smartphones, laptops and catalytic converters. The 2010 Nobel Prize for Chemistry went to three scientists – Richard Heck and Ei-ichi Negishi from the USA, and Akira Suzuki from Japan – who had developed a process of carbon-to-carbon bonding reliant on a palladium catalyst.

▲ Precious silver also has many practical and industrial uses.

▲ The Tollens reaction coats a glass surface with silver, an experiment that students can do in the laboratory.

Fast Facts

SILVER [Ag]
ATOMIC NUMBER 47

Character: Silver has the highest conductivity for heat and electricity of any metal, and the highest reflectivity. Silvered mirrors were perfected by German chemist Justus von Liebig in 1856.

Discovery: Our ancient ancestors were just like us in liking shiny silver trinkets, and people have been mining, extracting and working silver for at least 5,000 years. The landscapes of present-day Iraq, Turkey and Greece all bear the tell-tale signs of industrial-scale silver mining and refining from around 3000 BCE. The ancient civilizations of South America were also familiar with silver extraction and smelting techniques.

Name: The English word silver stems from the Old English word *siolfor*, which is in turn a corruption of the Old Norse word *silfr*. Its chemical symbol Ag is from the Latin word *argentums*, which also gives Argentina its name.

World sources: Mexico, Peru, China, Bolivia, Chile, USA, Peru, Turkey, Australia, Iran, Norway, Russia, Honduras, Germany and Canada all have silver ore deposits. Silver is important in electronics because of its conductivity, and in medicine because of its low chemical reactivity and antimicrobial properties.

Fast Facts

CADMIUM [Cd]

ATOMIC NUMBER 48

Character: Cadmium is a malleable and highly toxic silvery-blue metal.

Discovery: In Hanover, Germany, during the 1880s, the city's pharmacists discovered a problem. When they made their skin ointments by cooking up calamine to produce zinc oxide, the process sometimes yielded an unpleasant smelly yellowish by-product. The Inspector of Pharmacy, Friedrich Stromeyer of Göttingen University, could not identify the substance responsible – and therefore realized he had found a new element.

Name: Cadmium derives from the Latin word *cadmia*, meaning calamine.

World sources: Several countries have cadmium production capacity, but it is being phased out of its major uses such as battery manufacture because it is poisonous.

▲ *Van Gogh's* Sunflowers.

CADMIUM COLOURS

Vincent van Gogh painted his *Sunflowers* using lead chromate pigments known as chrome yellows. But 19th-century artists soon realized with dismay that the chrome yellows were not stable to light. Claude Monet used cadmium yellow in his *Waterlilies* and *Irises*, believing the pricier pigment would help his work last longer. While it is toxic, cadmium can produce dazzling colours all the way from pale and golden yellow, through orange, to red and maroon.

Yellow is just one of colourful cadmium's tones. ▶

▲ *Nanocrystals of cadmium selenide may find applications in a new generation of LED screens.*

Fast Facts

HAFNIUM [Hf]

ATOMIC NUMBER 72

Character: Hafnium is a bright and reflective silvery metal, strongly resistant to corrosion.

Discovery: In 1911, French chemist Georges Urbain announced that he had discovered the element that belonged in position 72 in the periodic table, calling it celtium. But it turned out Urbain had merely rediscovered an existing element: lutetium. In 1923, Hungarian chemist George Charles de Hevesy, working in partnership with Dutch physicist Dirk Coster, found an intriguing trace substance in a lump of zirconium ore. To extract the mystery material from zirconium ore was decidedly tricky, but when eventually isolated in 1925, it turned out to be element 72.

Name: Hafnium takes its name from the Latin word for Denmark's capital city Copenhagen. Although neither of the element's discoverers were Danish, they undertook their breakthrough work on hafnium while in Copenhagen.

World sources: France, USA, Australia, Brazil and Malawi have deposits of hafnium, often in combination with zircon ore. Annual production is only around 5,000 tonnes currently, but this could rise if demand increases in the electronics industry.

▲ Hafnium deposits often appear in combination with zirco ore. In this mineral sample are green zircon crystals.

▲ Tantalum conducts electricity and heat very well, and forms components in many portable electronic devices.

▲ *In Greek mythology, Tantalus could not reach anything to eat or drink as a punishment.*

▼ *Coltan ore contains tantalum and niobium metals, in demand for electronics manufacture.*

Fast Facts
TANTALUM [Ta]
ATOMIC NUMBER 73

Character: Tantalum is a very hard, light grey metal, and only two other metals – tungsten and rhenium – have higher melting points.

Discovery: As early as 1802, Swedish chemist Anders Gustav Ekeberg extracted what he believed was a new element. However, British scientist William Hyde Wollaston, who had discovered palladium, thought Ekeberg was seeing niobium. Decades later, German mineralogist Heinrich Rose separated tantalum and niobium – it turned out they are often entangled together in ore samples – and showed they were different elements. Pure tantalum was finally isolated by German chemist Werner von Bolton in 1903.

Name: Tantalum's name derives from Tantalus, who was King of Sisyphus in Lydia. One day, Tantalus stole the food of the Greek gods, and his punishment was to be made eternally hungry and thirsty and unable to reach any food or drink. The discovery of the element was similarly tortuous.

World sources: Tantalum is a conflict mineral, meaning that its production takes place without worker rights, in an area of armed conflict, and is sold illicitly to fund the fighting, for example in the Democratic Republic of Congo. Demand for the metal is high because of its use in capacitors for smartphones.

Fast Facts
TUNGSTEN [W]
ATOMIC NUMBER 74

Character: Tungsten is an extremely hard and dense greyish-white metal, with one of the highest boiling points of all the elements.

Discovery: It took three goes to find tungsten. In 1779, Peter Woulfe suspected a Swedish mineral contained a new metal. Two years later, Carl Wilhelm Scheele managed to isolate an oxide from the same mineral. Finally, Spanish brothers Fausto and Juan José Elhuyar reduced the oxide to reveal element 74.

Name: The element goes by the name of tungsten, from the mineral name in Swedish – *tung sten*, meaning heavy stone. Its alternative name, wolfram, comes from the tungsten ore mineral wolframite.

World sources: China and Russia are the top tungsten producers.

▶ *Tungsten carbide tips on drill bits make them tough enough to tackle concrete, brick and stone.*

▲ *Thomas Edison tested thousands of materials looking for a suitable filament for his light bulb, carrying out the first successful experiment in 1879. In 1904, Hungarian inventors patented a lamp with a tungsten filament that gave brighter light.*

▲ *Osmiridium, an alloy of osmium and iridium, makes strong fountain pen nibs.*

Fast Facts
OSMIUM Os
ATOMIC NUMBER 76

Character: Osmium is an extremely hard, yet brittle, metal. Bright grey in colour with a distinctively bluish tint, osmium emits a potent smell. It is the densest stable element, approximately twice the density of lead.

Discovery: English chemist Smithson Tennant investigated platinum ore alongside his friend and collaborator William Hyde Wollaston, who discovered rhodium and palladium. In 1803, Tennant managed to separate from his platinum sample two new elements, iridium and osmium.

Name: Tennant named osmium using the Greek word for smell, in recognition of its unpleasant aroma.

World sources: South Africa, Russia and Canada have the largest reserves of osmium. However, world production is modest. When Thomas Edison was looking for a filament material with a high melting point for his light bulbs, he considered osmium, but tungsten became the favoured choice.

Fast Facts
RHENIUM Re
ATOMIC NUMBER 75

Character: A dense, heavy, bright silvery metal, rhenium has a very high boiling point.

Discovery: In 1905, Japanese chemist Masataka Ogawa announced he had discovered element 43. However, his finding could not be replicated, and was forgotten. Ogawa died in 1930 – but research in the 1990s showed that what he had actually found was element 75. This element had been discovered in 1925 by Ida Tacke, Walter Noddack and Otto Berg, who processed over half a tonne of ore to produce a single gram.

Name: The name rhenium derived from *Rhenus*, the Latin name for the German Rhine river.

World sources: Chile, Poland, Russia and Peru produce rhenium.

▶ *The densest natural element, osmium is a hard, brittle metal in the platinum family.*

IRIDIUM: THE CLUE TO THE DINOSAURS' DEMISE

What wiped out the dinosaurs? A leading theory is that, 66 million years ago, an asteroid 10–15km (6–9 miles) in diameter hit the Earth. The colossal rock had a devastating effect.

Fire raged, fed by higher oxygen levels in the atmosphere than today, and giant tidal waves set out across the seas. The asteroid impact sent a vast, dense dust cloud into the air, blocking out the sunlight and causing an impact winter. Photosynthesis ground to a halt, and with the death of green plants, so went the animals that ate them – as well as the animals that ate those animals. Three-quarters of plant and animal species became extinct.

The crucial early evidence for an asteroid impact was gathered by a physicist, Luis Alvarez, his geologist son, Walter Alvarez, and two chemists, Frank Asaro and Helen Vaughn Michel. It came from a layer of rock found all over the world, called the Cretaceous-Paleogene (K-Pg) boundary. In 1980, the team discovered that this thin sedimentary layer contains levels of iridium dozens of times higher than normal in the Earth's crust.

▲ *Walter and Luis Alvarez.*

Iridium as an element is rare in our daily existence. Its tendency to bond with iron means that early deposits disappeared into the planet's core when the Earth's layers were being formed. However, iridium is common in most

◀ *A crater in the Yucatan Peninsula in Mexico provides evidence of an asteroid impact.*

UNDERGROUND EVIDENCE OF THE CHICXULUB CRATER

Geophysicists Antonio Camargo and Glen Penfield carried out a magnetic survey of the Yucatan Peninsula (coastline shown in white) in 1978, and discovered a large underground arc. In the 1990s, samples confirmed it was a huge buried crater.

asteroids and comets – so the team proposed this as evidence for an impact with an extra-terrestrial rock.

Then, in the 1990s, geologists uncovered evidence of a giant crater, 180km (112 miles) in width, buried underneath the Yucatan Peninsula in Mexico. It dated to the late Cretaceous Period – so that was a match. There was also evidence of ancient tidal wave damage in nearby Texas, and a thick deposit of glassy spheres in Mexico, potential evidence of a massive impact nearby.

Scientists still debate whether an asteroid impact was the only factor that led to the dinosaurs' departure, but the Earth's iridium-rich layer gives us a tantalizing view into the Earth's distant past.

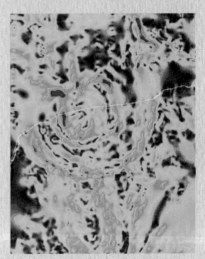

▲ *An iridium-rich layer of rock all over the world led to the confirmation of an asteroid impact 66 million years ago.*

Fast Facts

IRIDIUM [Ir]
ATOMIC NUMBER 77

Character: Iridium is an exceptionally hard metal and the second densest, after osmium.

Discovery: In his investigations of platinum ore in 1803, chemist Smithson Tennant separated a substance that he found created a rainbow of compounds. It was a new element – the second he had discovered the same year, along with osmium.

Name: Tennant called the colourful element iridium, which he derived from Iris, the Greek goddess of the rainbow.

World sources: Iridium is one of the least common stable elements in the Earth's crust. South Africa, Russia and Canada have deposits, and the metal is also found combined with other platinum group elements.

THE PLATINUM GROUP

On Christmas Eve in 1800, chemist William Hyde Wollaston received the best possible gift from fellow scientist Smithson Tennant. It was a sample of platinum ore which Tennant had bought, and the two men agreed to investigate. Between 1802 and 1805, they discovered four new elements in the ore – rhodium, osmium, iridium and palladium – and Wollaston developed a method for making platinum malleable – a finding that made him rich.

Along with ruthenium, added to the fold in 1844, the platinum group is a cluster of six precious, corrosion-resistant elements that sits in the middle of the periodic table. Rare though they are, their applications make them valuable today – in medical implants, catalytic converters, fuel cells, electronic devices and displays, and as catalysts in the chemical industry.

◄Smithson Tennant named iridium after Iris, the rainbow goddess, because of its range of coloured of compounds.

Fast Facts
PLATINUM [Pt]
ATOMIC NUMBER 78

Character: One of the densest metals, platinum is shiny, silvery-white and precious. It is highly nonreactive and stable, and in an alloy with 10 per cent iridium it forms the international standard kilogram weight, stored in Saint-Cloud, France.

Discovery: Civilizations in pre-Columbian South America found and used platinum, but it took two centuries for colonists to recognize its potential value. In 2007, German scientist Gerhard Ertl won the Nobel Prize for Chemistry. His discoveries included how platinum provides the right surface for the chemistry inside catalytic converters in car exhausts.

Name: Platinum's name comes from the Spanish word *platina*, meaning little silver.

World sources: South Africa has the largest platinum deposits today, and the largest production capacity.

▲ *Clockwise from top left: osmium, ruthenium, rhodium, palladium, platinum and iridium.*

◄►
The platinum-based international kilogram weight (left), and an even more accurate sphere of silicon (right) that will replace it.

THE PLATINUM AGE

The largest single recorded delivery of platinum from Colombia to colonial Spain came by ship in 1789. There were 1,630kg (3,594lb) of the metal's ore on board, in 41 crates. Accompanying the platinum were ten sacks of cacao beans, and a jaguar, to amuse the king.

In the 1500s, when Europeans arrived in Colombia, they discovered the Muisca civilization in control of a thriving industry in gold and emeralds. The legend of *El Dorado* grew up out of stories about a chieftain covered in gold dust or even an entire lost city of glittering wealth.

Under colonial rule, mines worked by slaves fed the greedy European appetite for gold. But while there was plenty of the golden metal, it was often contaminated by another substance that was silvery, dense, and hard to melt. It would creep into gold ingots, either in error or as a makeweight, causing problems for the royal mint in Bogota. The rogue metal was platinum.

Earlier peoples in the region had turned platinum into intricate masks and jewellery. But for the Europeans, the white metal was an annoyance. They saw *platina* as an unripe form of gold, worth less than iron. By the 1700s,

◄►
Gold artefacts in Colombia led to the legend of El Dorado. The Muisca people used gold figurines as part of offerings.

With a monopoly on the supply of this exciting new metal, the king of Spain appointed a chemist, Pierre-François Chabaneau, to investigate platinum's properties. In 1785, Chabaneau announced he could produce malleable platinum in large quantities. A great chalice was crafted from platinum and presented to the Pope. The Spanish age of platinum had begun.

miners and merchants could be fined for failing to remove it from gold bars. Once extracted, the worthless platinum was dumped in rivers.

In 1748, a Spanish explorer called Antonio de Ulloa wrote about the Colombian metal and took samples back to Spain to study. In 1750, English scientist William Brownrigg presented an account of the new substance to the Royal Society in London. News spread of platinum, the eighth metal discovered, joining the seven known since classical times: gold, silver, mercury, copper, iron, tin and lead.

◄ *Once considered worthless by European colonists, platinum became so prized that the Pope was given a platinum chalice.*

Fast Facts

GOLD [Au]

ATOMIC NUMBER 79

Character: Heavy yet soft and inert, gold is a precious metal that is malleable, much-coveted and retains its glowing colour over prolonged periods of time.

Discovery: Humans have been in love with gold since prehistoric times. Graves in Iraq dating back 5,000 years contained gold jewellery. The Ancient Egyptians had already begun gold-mining activities 4,000 years ago in Nubia. Pure gold coins were minted during the reign of King Croesus of Lydia in modern-day Turkey around 550 BCE.

Name: Our modern word gold is a direct import from the same word in old Anglo-Saxon, which probably originated from the Old Norse word *gull*. Oddly, gold – the undoubted star of the transition metals – is not recognized by a chemical symbol G, but by Au, from the Latin word for gold: *aurum*.

World sources: South Africa, China, USA, Australia, Peru, Russia, Indonesia and Canada are the world's main suppliers. Such is its appeal that around 90 countries produce gold commercially.

▲ *Gold holds a strange fascination for many people.*

▲ *A pair of gold bracelets form part of the treasure of Lydia in Turkey.*

Fast Facts

MERCURY Hg
ATOMIC NUMBER 80

Character: Shiny, silvery-grey mercury is the only metal that is a liquid at room temperature.

Discovery: Mercury is one of the original eight metals known since ancient times. Cave paintings made in France and Spain 30,000 years ago, during the Paleolithic, used mercury in a red pigment called cinnabar. For centuries, some of science's biggest names were convinced that alchemy was possible, and that you could transmute elements into gold if you could find the right ingredients to add to a mercury base. Isaac Newton carried out many alchemical experiments, and a lock of his hair revealed excessive levels of mercury, which is toxic.

Name: The fleet-footed messenger Mercury gives his name to this element, which is also sometimes called quicksilver. The chemical symbol Hg is from a former name hydrargyrum, meaning water-silver.

World sources: China and Kyrgyzstan are among the current producers of mercury. Some earlier sources in Europe and the Americas have been exhausted.

▼ *Mercury takes its chemical symbol, Hg, from a word meaning water-silver.*

Fast Facts
RUTHERFORDIUM [Rf]

ATOMIC NUMBER 10

▲ *American nuclear chemist James Harris (seen here with Albert Ghiorso) played a vital role in the discovery of elements 104 and 105, designing and purifying the targets for bombardment in the particle accelerator.*

Character: Rutherfordium was the first element discovered in the super-heavy category, with 104 or more protons in the nucleus. It is radioactive and short-lived.

Discovery: Element 104 first made an appearance in 1964 when a team at the Joint Institute for Nuclear Research (JINR) in Dubna, Russia, bombarded plutonium-242 with neon-22. They confirmed their finding in 1966 and proposed the name kurchatovium after a leading Soviet physicist Igor Kurchatov. In 1969, a team at the Lawrence Berkeley Laboratory in California also made the element and suggested naming it after nuclear pioneer Ernest Rutherford.

Name: Both Russian and American scientific teams are credited with discovering element 104, which international bodies decided in 1992 would be called rutherfordium.

World sources: Rutherfordium only exists in the laboratory.

◀ *The Joint Institute for Nuclear Research (JINR) in Dubna, Russia, is an international research centre for nuclear sciences and elementary discovery.*

Fast Facts
DUBNIUM [Db]
ATOMIC NUMBER 105

Character: Discovered in 1968, dubnium is a super-heavy element with 12 isotopes.

Discovery: Scientists at the Joint Institute for Nuclear Research (JINR) in Dubna, Russia, created this element at around the same time it was also made at the Lawrence Berkeley Laboratory in the United States.

Name: The naming of element 105 caused controversy, with the Russian team suggesting it should honour Niels Bohr, and the Americans preferring Otto Hahn. In the end, the name bohrium went to element 107, and element 105 became dubnium in recognition of the contribution of JINR.

World sources: Dubnium is radioactive and short-lived, existing solely in the laboratory.

Fast Facts
SEABORGIUM [Sg]
ATOMIC NUMBER 106

Character: Scientists predict this element would be a silvery metal, but only a few atoms have ever been made, which quickly decayed.

Discovery: In the rush to find more super-heavy elements during the 1970s, an American team created an isotope of element 106. Russian scientists then made two further isotopes.

Name: Seaborgium is named after Glenn Seaborg, who was involved in discovering several elements heavier than uranium. It was unusual to name an element after a living scientist, and the name was adopted among a number of hard-argued names for elements 104–109.

World sources: Like its near neighbours in the periodic table, seaborgium must be made by bombarding elements together.

◄ *Glenn Seaborg had the experience of having an element named after him during his lifetime.*

The final six elements of the transition metals are created in tiny quantities in the laboratory, but all fill predicted spaces in the periodic table.

Fast Facts
BOHRIUM [Bh]
ATOMIC NUMBER 107

Character: Bohrium is highly radioactive with a half-life of 61 seconds.

Discovery: Element 107 was confirmed at the Helmholtz Centre for Heavy Ion Research in Darmstadt, Germany, in 1981, the first of five super-heavy elements spotted there.

Name: Scientists suggested nielsbohrium for the element's name. Bohrium was accepted in 1997.

Fast Facts
HASSIUM [Hs]
ATOMIC NUMBER 108

Character: Extremely radioactive, metallic hassium only lasts a few seconds.

Discovery: Hassium is a child of the 1980s. A team in Darmstadt, Germany, bombarded a target made of lead with atoms of iron, announcing element 108 in 1984.

Name: The word hassium derives from the name of the German state Hesse, where the experiment, led by Peter Armbruster and Gottfried Münzenberg, took place.

Fast Facts
MEITNERIUM [Mt]
ATOMIC NUMBER 109

Character: It is possible that meitnerium might be the densest of all elements.

Discovery: In a week-long experiment in 1982 at the Helmholtz Centre for Heavy Ion Research in Darmstadt, the team found one atom of element 109 which decayed after five milliseconds. Russian scientists later confirmed the finding.

Name: The name honours nuclear fission pioneer Lise Meitner.

Fast Facts
DARMSTADTIUM `Ds`
ATOMIC NUMBER 110

Character: Possibly similar to platinum, element 110 has a half-life of microseconds.

Discovery: Scientists in Darmstadt, at the Helmholtz Centre for Heavy Ion Research, used nickel to bombard a lead target and fused together an element with 110 protons in the nucleus.

Name: Darmstadtium is named after the location of its discovery.

Fast Facts
COPERNICIUM `Cn`
ATOMIC NUMBER 112

Character: Copernicium may be more like a gas than a metal – but no one is quite sure.

Discovery: Created from particles of zinc bombarding a lead target, element 112 was discovered in 1996.

Name: The team led by Sigurd Hofmann in Darmstadt, Germany, suggested the name to honour Polish scholar Nicolaus Copernicus.

Fast Facts
ROENTGENIUM `Rg`
ATOMIC NUMBER 111

Character: In the same column of the periodic table as copper, silver and gold, element 111 may share some of their characteristics.

Discovery: In 1994, the Helmholtz Centre in Darmstadt, Germany, was party to the birth of another new element. The RIKEN Linear Accelerator in Japan confirmed the finding in 2003.

Name: Element 111 honours Wilhelm Roentgen, who discovered X-rays.

▲ 16th-century scientist Nicolaus Copernicus is remembered in the name of element 112.

◀ Wilhelm Roentgen discovered X-rays in 1895, and became the recipient of the first Nobel Prize for Physics in 1901.

PERIODIC PIONEERS
LISE MEITNER

Lise Meitner saw the study of physics as a battle for ultimate truth – and it was not the only struggle she encountered in her long life. She was born in Austria in 1878, in a period when girls officially left school at 14. Nonetheless, she entered the University of Vienna in 1901. Meitner studied under Ludwig Boltzmann and gained a Ph.D in physics in 1906.

▲ *Lise Meitner, born in 1878 in Austria, studied physics and co-discovered the process of nuclear fission.*

She hoped to work with Marie Curie, but as no roles were available, persuaded Max Planck to allow her to attend the University of Berlin where women were usually barred from his lectures.

Meitner began publishing scientific papers, and formed a working partnership with chemist Otto Hahn, studying radioactivity. The research continued, despite First World War service, and in 1917 Meitner and Hahn co-discovered the element protactinium. By 1926, Meitner was the first female professor of physics in Germany.

The Second World War brought hostility from the Nazis toward Meitner, whose family was Jewish. Her nephew, physicist Otto Frisch, had already had to leave Germany, along with many other scientists. Meitner fled to Sweden in 1938.

Shortly afterward, Frisch was visiting Meitner in Sweden when Otto Hahn wrote with puzzling experimental results. A collision of a neutron with a uranium nucleus had apparently made it burst apart, producing a much lighter element, barium. Hahn hoped Meitner could find a theoretical explanation.

Aunt and nephew put together the mathematics of the reaction and realized that the heavier atom was splitting into lighter fragments, a process they named fission. Meitner also saw the huge scale of the energy that such a split would release.

▲ Lise Meitner with Otto Hahn.

This was new fundamental science – and the physics behind the atomic bomb, although Meitner refused to have any part in the Manhattan Project when she was later asked to join the team.

Otto Hahn published the chemistry of fission in 1939. He made no reference to Meitner because he was still living in Nazi Germany, and in 1944 he won the Nobel Prize for Chemistry. For many people, Meitner's exclusion has been controversial ever since.

However, Meitner's contribution in the battle for ultimate truth was honoured in 1997. In an extraordinary accolade given to so few, element 109 was named meitnerium.

NUCLEAR FISSION

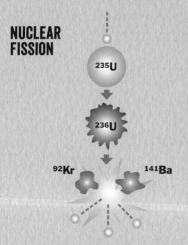

^{235}U

^{236}U

^{92}Kr ^{141}Ba

▲ Uranium atoms in Otto Hahn's experiment split to form two lighter atoms, krypton and barium, plus neutrons and a huge amount of energy.

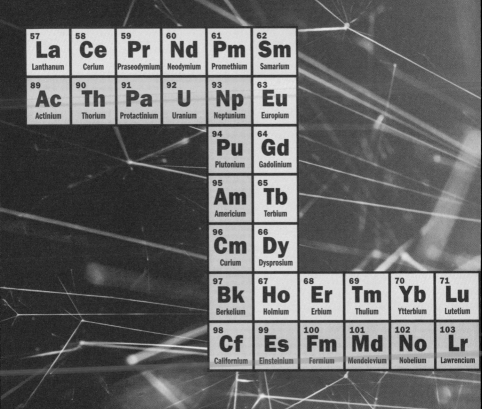

CHAPTER 6

LANTHANIDES AND ACTINIDES

WHICH OF THE LANTHANIDES REALLY MAKES SPARKS FLY?

WHY WAS IT SO PERFECT THAT NASA'S NEW HORIZONS PROBE WAS POWERED BY NEPTUNIUM?

WHO DISCOVERED THE MOST ELEMENTS IN THE PERIODIC TABLE?

LANTHANIDES AND ACTINIDES

The periodic table is usually conveniently arranged into 18 columns. But you may have noticed a blip between barium and hafnium, and also between radium and rutherfordium. At these points, there are two extra rows of elements to fit in to the sequence of atomic numbers.

The lanthanides and actinides – also sometimes known as lanthanoids and actinoids – are usually shown beneath the rest of the elements, simply because fitting them in by number order would make the overall table so wide. Lanthanum gives its name to the lanthanides, the upper series of elements ending in lutetium, while the row below, which starts with actinium and runs to lawrencium, contains the actinides.

As ever, electronic structure helps explain these elements' behaviour. But unusually, the lanthanides and actinides leave their outer electron arrangement unchanged, while gradually adding electrons to their inner f orbitals. This explains why they have such similarities in physical terms.

Car batteries and wind turbines (below and right) contain neodymium and praseodymium. ▼

▶ *Portable electronics use a range of lanthanides, with currently only a limited amount of recycling after the devices are thrown away.*

Their family resemblance made them hard to distinguish, and discovery of the lanthanides came in painfully slow stages. It began when a heavy, reddish mineral turned up in a Swedish mine. In 1803, chemists spotted that it contained a new element, cerium (element 58). In 1839, it turned out that the cerium ore also contained lanthanum (57) and something called didymium.

But in 1879, one type of didymium revealed itself to contain samarium (62) and another type, in 1885, to contain praseodymium (59) and neodymium (60). Samarium later yielded gadolinium (64) and europium (63).

In the meantime, a chunk of black mineral came to light in another mine in Sweden. This yielded terbium (65), erbium (68), ytterbium (70), lutetium (71), holmium (67), thulium (69) and dysprosium (66). Promethium, the missing element 61, emerged in 1945 from a nuclear reactor.

Actinides had a different discovery route. Thorium (90) was found in a mineral in 1829 and uranium (92) in 1841. Actinium (89) emerged in 1899 from uranium ore, while protactinium (91) was confirmed in 1918. The actinides from neptunium (93) to lawrencium (103) were all synthesized and do not exist in nature.

Fast Facts
LANTHANUM La
ATOMIC NUMBER 57

Character: Reactive in comparison with other lanthanides, lanthanum is soft enough to cut with a knife.

Discovery: Lanthanum was an unexpected discovery for Swedish scientist Carl Gustaf Mosander in 1839, who was in the process of investigating a compound of the next element, cerium.

Name: The Greek word *lanthano* means to escape notice, as lanthanum had done. It was only isolated in a pure form in 1923.

World sources: The mineral bastnaesite is a source of pure lanthanum, which is actually three times more common than lead.

▼ *In the film industry, lanthanum readily produces showers of sparks for special effects.*

Fast Facts
CERIUM Ce
ATOMIC NUMBER 58

Character: Cerium sparks when struck – but this is one of the pure metal's only uses. It corrodes easily and reacts with water.

Discovery: Cerium was isolated in 1875 when William Hillebrand and Thomas Norton separated it from molten cerium chloride. Nanoparticles of cerium oxide can make diesel fuel more efficient.

Name: Ceres ought to offer fertile applications – its name recalls an asteroid, Ceres, which is named after the Roman god of agriculture.

World sources: The most abundant of the lanthanides, cerium is straightforward to extract from minerals.

▲ *Cerium makes up nearly half of an alloy called mischmetal, used to strike sparks in cigarette lighters.*

Fast Facts
PRASEODYMIUM **Pr**
ATOMIC NUMBER 59

Character: Praseodymium is a soft and silvery metal that develops a green oxide coating in the air.

Discovery: When Carl Mosander separated a new substance from a cerium mineral, he thought it was a new element and called it didymium. However, Karl Auer von Welsbach found later that didymium contained two different elements, one of which was praseodymium, purified in 1931.

Name: The rather complicated name comes from the Greek *prasios didymos* meaning green twin, as its salts are a leek-green colour.

World sources: Praseodymium forms useful alloys for high-tech and magnetic applications, and is extracted from minerals, primarily monazite and bastnaesite.

Fast Facts
NEODYMIUM **Nd**
ATOMIC NUMBER 60

Character: Neodymium is physically similar to praseodymium. Its main use is as an alloy with boron and iron, discovered in 1983, that can make magnets strong enough to lift thousands of times their own weight.

Discovery: In 1885, Karl Auer von Welsbach discovered neodymium hidden within the substance called didymium along with praseodymium. He was a student of Robert Bunsen, the German chemist famous for his burner.

Name: Neodymium means new twin, from the Greek *neo didymos*.

World sources: Mines in China, USA and Brazil all produce minerals bearing neodymium.

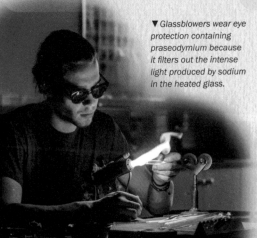

▼ *Glassblowers wear eye protection containing praseodymium because it filters out the intense light produced by sodium in the heated glass.*

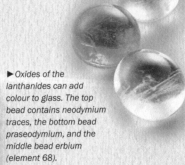

▶ *Oxides of the lanthanides can add colour to glass. The top bead contains neodymium traces, the bottom bead praseodymium, and the middle bead erbium (element 68).*

Fast Facts
PROMETHIUM [Pm]
ATOMIC NUMBER 61

Character: The rarest lanthanide element, promethium is a radioactive metal and source of X-rays.

Discovery: Promethium occurs in such minuscule amounts in nature that its existence was only confirmed when American scientists separated it from uranium products made in a nuclear reactor.

Name: Grace Coryell, who was married to one of the element's discoverers, suggested its name after Prometheus, who stole fire from the gods in Greek myth.

World sources: Useful promethium deposits do not occur naturally, but the element can be made in a nuclear reactor.

▼ In the past, watch faces were sometimes marked with glowing promethium paint, as were controls in the Apollo Lunar Module. It is now used in diving watches.

Fast Facts
SAMARIUM [Sm]
ATOMIC NUMBER 62

Character: A silvery-white metal, samarium is useful in making lasers and as an oxide added to glass and ceramics. Cancer treatments can involve the isotope samarium-153.

Discovery: In 1879, French chemist Paul-Émile Lecoq de Boisbaudran found samarium in a sample extracted from a mineral called samarskite.

Name: Samarium was named after the mineral samarskite, which in turn recalls a Russian mining engineer called Vasili Samarsky – the first person to have a chemical element named after them, albeit indirectly.

World sources: Today, samarium is extracted from monazite instead of samarskite.

▲ An alloy of samarium and cobalt makes a powerful magnet.

EUROPIUM [Eu]
ATOMIC NUMBER 63

Character: A silvery metal with a consistency like lead, europium reacts with water and in the air.

Discovery: Europium was another lanthanide found by French chemist Paul-Émile Lecoq de Boisbaudran, in 1890. It was isolated by Eugène-Anatole Demarcay.

Name: Demarcay's chosen name for the element was europium.

World sources: Minerals such as bastnaesite, monazite and xenotime contain europium.

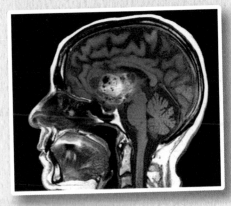

▲ *Gadolinium is a useful contrast medium for MRI scans, in this case revealing a brain tumour.*

▼ *Euro notes are protected against fraud by incorporating ink with europium salts that glow under UV light.*

GADOLINIUM [Gd]
ATOMIC NUMBER 64

Character: Soft enough to work, and usually mingled with other lanthanide metals, gadolinium is silvery-grey.

Discovery: Swiss chemist Jean Charles Galissard de Marignac separated an oxide of element 64 in 1880, and Paul-Émile Lecoq de Boisbaudran purified it in 1886.

Name: Gadolinium takes its name from the mineral gadolinia in which it occurs. Johan Gadolin was a Finnish chemist who discovered the element yttrium.

World sources: In Earth's crust gadolinium is as abundant as nickel and arsenic.

▼ *Terbium phosphors were once used in many cathode ray tube televisions.*

Character: Lustrous and silver, pure dysprosium can only be separated in the laboratory using modern techniques.

Discovery: It took 58 attempts, but Paul-Émile Lecoq de Boisbaudran finally managed to identify dysprosium in a mineral sample in 1886. He carried out his painstaking experiments on the marble slab of his home's fireplace.

Name: Boisbaudran called the element after the Greek word *dysprositos*, meaning hard to get.

World sources: Dysprosium is currently extracted from clay ores sourced in southern China. It has growing applications in magnets for wind turbines and car motors.

Character: Silvery-white and soft, terbium-based phosphors provide the green light which, combined with red and blue phosphors, makes high-efficiency white lighting.

Discovery: Carl Gustaf Mosander found terbium in 1843, but only as a yellow oxide. A process called ion exchange, invented in the 1950s, finally separated terbium in pure form.

Name: Terbium is one of four elements named after the village of Ytterby in Sweden.

World sources: Terbium occurs in many minerals, including cerite, gadolinite, monazite, xenotime and euxenite.

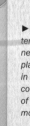

▶ *Dysprosium, terbium and neodymium play a role in electronic components of electric car motors.*

Fast Facts
HOLMIUM Ho
ATOMIC NUMBER 67

Character: Bright and silvery in its pure form, holmium has applications in nuclear reactors as an absorber of neutrons. It can also make the strongest magnets of any element in the periodic table.

Discovery: Marc Delafontaine and Louis Soret in Switzerland, and Per Teodor Cleve in Sweden, discovered holmium at about the same time in 1878.

Name: The name holmium came from the Latin word for Stockholm, the capital of Sweden.

World sources: Holmium deposits exist in China, USA, Brazil and India.

▼ Erbium doping in optical fibres amplifies the signals over longer distances.

Fast Facts
ERBIUM Er
ATOMIC NUMBER 68

Character: Erbium gives glass a pink tint and improves the malleability of some alloys. It is a soft, silvery metal that tarnishes in the air.

Discovery: Pure erbium metal was not extracted until 1934, but chemist Carl Gustaf Mosander had found pink-hued erbium oxide in 1794, 140 years earlier.

Name: Erbium is one of the four elements named after Ytterby, the village where the mineral that contained them was unearthed.

World sources: By modern chemical methods, erbium can be extracted from monazite and bastnaesite minerals.

Fast Facts
THULIUM [Tm]
ATOMIC NUMBER 69

Character: The least abundant of the naturally occurring lanthanides, thulium is bright and silvery in its pure form.

Discovery: Per Teodor Cleve, a Swedish polymath of the 19th century, realized in 1879 that thulium was lurking in a sample of erbium.

Name: Thule is an ancient name for Scandinavia, and the source of this element's name.

World sources: Thulium ore is most common in China.

Thulium. ▶

Fast Facts
YTTERBIUM [Yb]
ATOMIC NUMBER 70

Character: Soft and silvery, ytterbium is more reactive than most of the lanthanides and is stored in sealed jars under vacuum or inert gas.

Discovery: Jean Charles Galissard de Marignac found ytterbium in 1878, but nobody could obtain a pure sample until 1953.

Name: Ytterbium is the fourth and final element named after the Swedish village of Ytterby.

World sources: Only about 50 tonnes of ytterbium is extracted each year, as so far it has limited applications.

Fast Facts
LUTETIUM [Lu]
ATOMIC NUMBER 71

Character: Lutetium is highly reactive and prone to catching alight. It is the hardest and densest of the lanthanides.

Discovery: French chemist Georges Urbain discovered lutetium in 1907, winning the race against researchers in Germany and the USA.

Name: The Romans called Paris *Lutetia* and this is where lutetium's name comes from.

World sources: Most of the lutetium we need is extracted in China – but this only amounts to 10 tonnes per year.

Fast Facts
ACTINIUM [Ac]
ATOMIC NUMBER 89

Character: First of the actinide series, actinium is silvery-white and soft, oxidizing quickly in the air. It glows blue because of its intense radioactivity.

Discovery: In 1899 French chemist André-Louis Debierne separated element 89 from leftover uranium ore from which Marie and Pierre Curie had extracted radium. Friedrich Oskar Giesel purified it in 1902.

Name: The name actinium comes from the Greek word *actinos*, meaning a ray.

World sources: Actinium occurs only in trace amounts.

Fast Facts
THORIUM [Th]
ATOMIC NUMBER 90

Character: Thorium is a silvery metal that is radioactive, and can be used as a fuel in fission reactors. It was the second radioactive element ever discovered, after uranium.

Discovery: Swedish chemist Jacob Berzelius found thorium in a rock sample in 1829 and realized it was a new element.

Name: Thorium is named after Thor, the god of war in Norse mythology.

World sources: There are few applications for thorium and it is a by-product of extracting other minerals.

Jacob Berzelius discovered the elements thorium, cerium and selenium. ▲

▲ *Thorium is more abundant than uranium and is a promising fuel for nuclear reactors, particularly in India where thorium reserves exist.*

Fast Facts
PROTACTINIUM `Pa`
ATOMIC NUMBER 91

Character: Radioactive protactinium is a silvery metal.

Discovery: In 1900, English physicist William Crookes separated something highly radioactive from uranium. Then in Germany, Kasimir Fajans and Otto Göhring showed that the unidentified substance rapidly decayed by beta emission.

Name: Otto Hahn and Lise Meitner chose the name proto-actinium when they discovered a longer-lasting isotope of the same substance, leading to the name protactinium.

World sources: The longest-lasting and most abundant isotope is a product of the decay of uranium-235.

▼ *Yellow and yellow-green uranium glass was most popular from the 1880s to the 1920s.*

Fast Facts
URANIUM `U`
ATOMIC NUMBER 92

Character: Uranium is the heaviest element in nature. Uranium-235 is the only natural substance that can sustain a chain reaction in a fission reactor.

Discovery: German chemist Martin Heinrich Klaproth realized he had found an oxide of a new element in 1789, and Eugène Péligot separated the first uranium metal in 1841. In 1938, Lise Meitner and Otto Frisch calculated that no stable elements exist beyond uranium, because the electrical repulsion between so many protons overcame the usual nuclear forces holding them together.

Name: Klaproth named the new substance after the planet Uranus.

World sources: Kazakhstan mines the most uranium, followed by Canada, Australia, Niger, Namibia and Russia.

Under UV light, uranium glass vessels glow green. ▶

Fast Facts
PLUTONIUM `Pu`
ATOMIC NUMBER 94

Character: Plutonium is silvery-grey, metallic and radioactive. Plutonium-238 pellets helped power NASA's New Horizons probe which, appropriately, flew by Pluto in 2015 as the first spacecraft to explore the dwarf planet.

Discovery: Glenn Seaborg and a team at the University of California, Berkeley, first made plutonium in 1940 during the Manhattan Project, but kept it secret because of the war. Plutonium isotopes can sustain a fission chain reaction, releasing huge amounts of energy. The element formed the core of the atomic bomb dropped on Nagasaki in 1945.

Name: Plutonium followed the pattern of naming elements after planets, although Pluto has since been demoted from planetary status.

World sources: While there are tiny traces of plutonium on Earth, it is primarily a synthesized element.

Fast Facts
NEPTUNIUM `Np`
ATOMIC NUMBER 93

Character: Radioactive and poisonous, neptunium can burst spontaneously into flames.

Discovery: In 1940, Edwin McMillan and Philip Abelson synthesized neptunium at the Lawrence Berkeley National Laboratory, as it is called today.

Name: The element's discoverers proposed neptunium for element 93, because it was next door to uranium, just as the planets neighbour each other in the solar system.

World sources: Neptunium is produced in nuclear reactions.

► *A fission bomb dropped on Nagasaki on 9 August 1945 released the equivalent of 20,000 tonnes of high explosive via a nuclear chain reaction in its plutonium core.*

177

Fast Facts
AMERICIUM [Am]
ATOMIC NUMBER 95

Character: Americium is shiny, silvery and radioactive, but has a place in many households as part of smoke alarms.

Discovery: Albert Ghiorso takes the prize for having had a hand in discovering the most elements – 12. He, Glenn Seaborg and other scientists produced element 95 at the University of Chicago in 1944, by bombarding plutonium with neutrons.

Name: The name americium sits below the lanthanide europium, and followed the geographical trend of names.

World sources: Americium once existed on Earth, but all natural sources have radioactively decayed.

Fast Facts
CURIUM [Cm]
ATOMIC NUMBER 96

Character: A silvery metal, curium is brittle and tarnishes in the air.

Discovery: Glenn Seaborg, Albert Ghiorso and team bombarded freshly discovered plutonium with alpha particles at the University of California, Berkeley, to make curium.

Name: The name of element 96 honours pioneers of radioactivity Marie and Pierre Curie.

World sources: Minuscule amounts of curium may exist in uranium deposits, and only a few grams are produced to meet demand each year.

▶ *Albert Ghiorso worked on early cyclotron experiments and is credited with co-discovering 12 new elements.*

▲ *Tiny amounts of americium are present in many homes worldwide inside ionization smoke detectors.*

▼ The Sather Tower overlooks the University of California, Berkeley campus, where early cyclotron experiments to find heavy elements took place.

Fast Facts
BERKELIUM [Bk]
ATOMIC NUMBER 97

Character: Like its neighbours, berkelium is silvery and radioactive.

Discovery: The discovery of element 97 relied on another recent finding, americium. Albert Ghiorso, Glenn Seaborg and Stanley Thompson bombarded it with helium nuclei to create a few atoms of the new substance in 1949.

Name: The name berkelium marked the location of the laboratory where the new element was made.

World sources: Berkelium is the least abundant of all naturally occurring substances, and vanishingly small amounts have ever been made.

Fast Facts
CALIFORNIUM [Cf]
ATOMIC NUMBER 98

Character: Californium is a radioactive metal that emits neutrons.

Discovery: Researchers Albert Ghiorso, Glenn Seaborg and a team at the University of California, Berkeley, once again made element 98. They bombarded curium with alpha particles to add the two extra necessary protons.

Name: Californium is named after the university and the state where the element was made.

World sources: Traces of natural californium exist on Earth, and fallout from nuclear testing has contributed small amounts to the environment.

Fast Facts
EINSTEINIUM [Es]

ATOMIC NUMBER 99

Character: A silvery, radioactive element, einsteinium is highly dangerous.

Discovery: Charred remains of coral from a secret American thermonuclear test zone in the South Pacific yielded the first sight of element 99, in 1952.

Name: Albert Ghiorso and his team at the University of California, Berkeley, discovered the element and named it einsteinium.

World sources: Today, if scientists need einsteinium, they bombard plutonium with neutrons.

▲ *The first hydrogen bomb detonated on a barge at Eniwetok Atoll in October 1952, with the same power as 10.4 million tonnes of high-explosive TNT.*

▲ *Italian-American physicist Enrico Fermi, after whom fermium is named.*

Fast Facts
FERMIUM [Fm]

ATOMIC NUMBER 100

Character: No one has seen enough fermium to be able to report on its appearance.

Discovery: Like einsteinium, element 100 also turned up in the ash and coral debris from the first hydrogen bomb test on the Eniwetok Atoll in the South Pacific.

Name: Fermium is named after Enrico Fermi, who won the Nobel Prize for Physics in 1938, and built the first nuclear reactor in 1942.

World sources: Fermium can be made by bombarding plutonium-239 with neutrons.

Fast Facts
MENDELEVIUM Md
ATOMIC NUMBER 101

Character: Mendelevium is probably a silvery metal, but from the quantity made it was not possible to tell.

Discovery: Element 101 was another discovery of Albert Ghiorso, Glenn Seaborg and colleagues at the University of California, Berkeley.

Name: The name mendelevium honours Dmitri Mendeleev, whose predictive ideas embodied in the periodic table led to so many discoveries.

World sources: None.

Fast Facts
NOBELIUM No
ATOMIC NUMBER 102

Character: Only a few atoms of nobelium have ever been made.

Discovery: For Albert Ghiorso, this is an element that got away. Four groups claimed the discovery of element 102, but officials eventually awarded the find to the Russian Joint Institute of Nuclear Research (JINR) in Dubna.

Name: Nobelium was the name given by Swedish researchers who thought they had found the element first, in 1957, and the name stuck.

World sources: None.

Fast Facts
LAWRENCIUM Lr
ATOMIC NUMBER 103

Character: Just a tiny amount of lawrencium has ever existed, and nobody knows what it looks like with the naked eye.

Discovery: Scientists in both Russia and America claimed priority for discovering lawrencium, and it was awarded to Albert Ghiorso and the team at Lawrence Berkeley Laboratory in the United States.

Name: Ernest Lawrence is honoured by the name of element 103. His atom-smasher, the cyclotron, was key to making and discovering numerous new elements.

World sources: None.

▲ *Ernest Lawrence invented the cyclotron in 1929, developed it into the particle accelerator, and won the 1939 Nobel Prize for Physics.*

PERIODIC PIONEERS
GLENN SEABORG

In 1951, when Glenn Seaborg won the Nobel Prize for Chemistry, it was for his work in extending the periodic table beyond element 92. Together with colleagues, he eventually synthesized a whole new row of heavier-than-uranium elements, sharing the prize with Edwin McMillan who created neptunium and contributed to many of the finds.

▲ *Maria Goeppert Mayer published ideas about a nuclear shell structure in 1950. Glenn Seaborg further proposed that certain numbers of protons and neutrons in the nucleus would create islands of stability and longer-lasting super-heavy elements.*

Seaborg was only 39 years old but had already had a busy career. In 1937, after gaining his Ph.D at the University of California, Berkeley, he worked as a personal laboratory assistant to Gilbert Lewis, a pioneering chemist whose contributions include the electron dot diagrams to represent atomic structures. Seaborg then conducted research into radioactivity using the university's Lawrence cyclotron.

In 1941, Seaborg was part of the team, with Emilio Segrè and Joseph Kennedy, who produced plutonium and showed that it underwent fission. They succeeded in isolating plutonium chemically, a key aspect of the Manhattan Project. In 1945, Seaborg signed a letter to President Truman asking him not to use the atomic bomb on Japan, alarmed at such "world-threatening peril", but to no avail.

In the late 1960s, Seaborg made another breakthrough in understanding super-heavy elements. His idea was based on a model of the nucleus in which the protons and neutrons are arranged in shells. Maria Goeppert Mayer had published the mathematics of nuclear shells in 1950, and she and J Hans D Jensen would share in the 1963 Nobel Prize for Physics for their work on the concept.

At certain magic numbers of protons and neutrons, theory predicted, the energy levels of the nuclear shells would be complete and this would give the particle a particularly long life – as in the case of an isotope of lead with 82 protons and 126 neutrons, for example. Seaborg suggested that there would be an island of stability among much heavier elements – around proton numbers 114, 120 and 126, all with specific magic numbers of neutrons.

Practice had not yet caught up with theory by the time Glenn Seaborg died in 1999. Russian scientists had come close by making atoms of flerovium (element 114) with 289 neutrons, in 1998 – but the total was 9 short of the stable 298. Maybe one day, as physics advances, Seaborg's predicted islands will feel like firmer ground.

▶ *Glenn Seaborg used this balance during his co-discovery of plutonium during the Second World War. He went on to find nine more elements.*

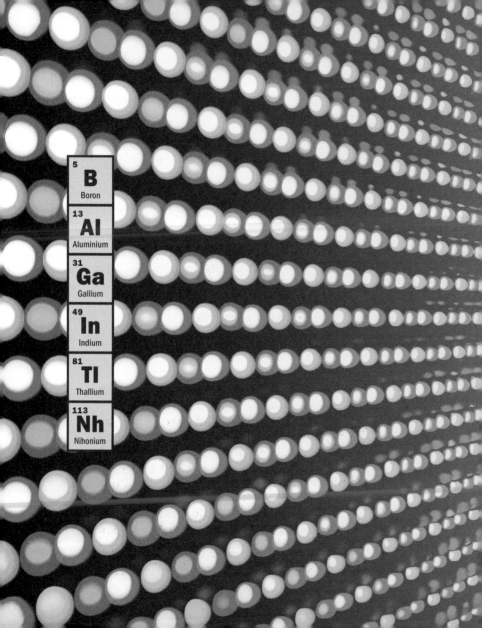

CHAPTER 7

THE BORON GROUP

How does boron keep the conversation flowing?

Which metal will melt in your hand?

What made nihonium such a wonderful
new year's gift for Japan?

THE BORON GROUP

After the practical uses and strength of the transition metals, the boron family seems like a group of more complex personalities. Boron itself refuses to conform to any easy categorization: it is a bit of an outlier from the group, being harder, and reacting only at much higher temperatures than fellow family members. All the other boron group elements are good conductors of electricity, but again boron sits out of this category as a metalloid, with properties of both metals and non-metals.

The group contains several post-transition metals – gallium, indium and thallium, and the artificially made nihonium. These elements are rather softer, weaker and have lower melting points than the more workaday transition metals.

The boron family of elements have three electrons in their outermost valence shell, and bond with other elements mostly through covalent bonds in which they share electrons. They all make oxides, in which two atoms bond with three atoms of oxygen, and can also all form stable compounds with the halogens – fluorine, chlorine and so on. None of them exists in pure form in nature.

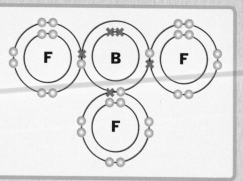

COVALENT BONDING

Boron (B) has three electrons in its outer shell, so it makes three shared pairs, in this case with fluorine (F), making boron trifluoride. These shared electrons are called covalent bonds.

WHAT MAKES METALS SHINY?

With the exception of boron itself, all the pure boron group are lustrous, silvery metals. The property of metallic lustre is related to the substances' atomic structure. Compared to non-metals, the outer electrons of metals are free to move around in a kind of sea surrounding positive nuclei from the metal atoms, and this is the key to their reflectiveness.

Whether an object is transparent, coloured or shiny depends on the way in which it interacts with incoming visible light. In a metal, the free electrons exist in a number of energy levels that match the energy of some of the incoming photons of light. The electrons can therefore absorb energy from the photons – but because they can freely vibrate, they then re-emit the light, giving a bright lustrous shimmer and reflectiveness.

▲ *Aluminium.*

◀ *Gallium.*

▲ *Indium.*

Boron is not as shiny as other metals in the group. ▶

[He] 2s^2 2p^2
4, 2, –4
77.2d 77 170

[He] 2
3
79.5

$_6$C

2.0
10^{-3}

$_5$B

[Ne] 3s^2

14 IVA

Fast Facts
BORON [B]
ATOMIC NUMBER 5

Character: The most common boron compound is disodium tetraborate decahydrate, known since ancient times as the detergent borax. It is extremely difficult to extract pure boron, but many compounds burn with a characteristic and beautiful green flame.

Discovery: In 1808, Joseph Louis Gay-Lussac and Louis-Jacques Thénard working in Paris, and Humphry Davy in London, first isolated boron and realized it was a new element.

▶ Borosilicate glass, which contains boric oxide, resists thermal expansion and is suitable for test tubes, oven-proof dishes and optical fibres.

Name: The name of the substance borax comes from its ancient Arabic name *buraq*. Humphry Davy saw a resemblance to carbon in the new substance and named it from a combination of borax and carbon.

World sources: Borax occurs naturally in large deposits in Turkey. Another large source is a mineral called rasorite, found in the Californian Mojave Desert.

▼ Boron burns with a green flame.

▲ Boron is useful for control rods in nuclear power stations because it absorbs neutrons, preventing an uncontrolled fission reaction.

▲ Narinder Singh Kapany pioneered fibre optics and gave them their name.

BORON AND THE COMMUNICATIONS REVOLUTION

More than 2 billion km (1.24 billion miles) of optical fibres crisscross the globe. Along these hair-thin fibres, an unimaginable quantity of information travels every second in the form of laser light signals. And much of it is shepherded to its destination using boron.

The term fibre optics was coined by Indian-born American physicist Narinder Singh Kapany. He first demonstrated successfully that he could send light through bent glass fibres, transmitting an image, during his doctoral work at Imperial College London, published in 1954.

In 1965, Charles Kao, who was born in China and worked in the UK, then discovered that the purity of the glass fibres was key to carrying light signals without loss – a finding that won him a part-share in the Nobel Prize for Physics in 2009.

During the 1970s, US scientists experimented with titanium and germanium to dope the core glass, and added boron to reduce the refractive index of the cladding. Erbium-doped glass has also improved the efficiency of amplification. Today, the world is encircled in fibre.

FIBRE OPTIC CABLES

Optical fibres work because the core has a higher refractive index – or ability to bend light – than the cladding. If the beam of light hits the interface between the two types of glass at an angle greater than the critical angle, it will not pass through into the cladding but reflects back, trapping the light safely in the core – a phenomenon called total internal reflection.

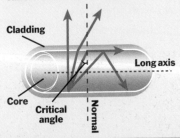

Cladding

Long axis

Core

Critical angle

Normal

Fast Facts
ALUMINIUM $\boxed{\text{Al}}$
ATOMIC NUMBER 13

Character: Silvery-white, soft and lightweight, aluminium is often used in alloys with stronger metals such as copper, magnesium and silicon. Its ability to conduct electricity and heat has transformed this corrosion-resistant metal into a commercial success.

Discovery: Danish physicist Hans Christian Ørsted was the first to make a sample of the element in 1825. Friedrich Wöhler, a German chemist, produced pure aluminium in 1827.

Name: Aluminium, or aluminum, comes from the name of alum, a common name for potassium aluminium sulfate used traditionally in the dyeing industry as a fixative. In Latin, *alumen* means "bitter salt".

World sources: Countries with reserves of aluminium minerals include China, Australia, Brazil and Jamaica. To extract aluminium by electrolysis from the molten ore mixture is energy-hungry, but once produced the metal can be recycled very efficiently.

▲ *Bauxite is the main ore of aluminium, the most abundant metal in Earth's crust. Iron oxide impurities give it its red colour.*

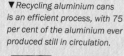

▼ *Recycling aluminium cans is an efficient process, with 75 per cent of the aluminium ever produced still in circulation.*

◄ *The Ottoman Empire had great expertise in weaving and dyeing, and supplied fixatives into Europe.*

HOW ALUM KICK-STARTED INDUSTRIAL CHEMISTRY

▲ Tolfa, Italy.

The medieval town of Tolfa, in Italy, has a colourful claim to fame. In 1461, its mountainous terrain yielded alunite, a source of alum – a vital fixative for dyes in the textile industry. Alum had previously been available in Europe only through imports from the Ottoman Empire. Now with a lucrative Italian source, the Pope took control of Tolfa's territory and began a long-lasting monopoly.

But things changed in the 1600s. The spotlight turned to North Yorkshire, UK, which has an altogether more pungent place in the story of alum. An English entrepreneur, Sir Thomas Chaloner, visited Rome and noticed that the geology and nature of the area resembled those near his Yorkshire home. He returned to England and set up alum production works that eventually spread across much of the local landscape.

ADDING URINE

The complex process involved extracting, then burning, huge piles of shale from nearby quarries for nine months. The result would be transferred into a pit for extraction of an aluminium sulfate liquor. Finally, human urine was added in the alum works – brought in by the tonne from as far away as Newcastle and London.

Smelly though it may have been, this alum-making was the first example of industrial-scale chemistry, and continued for 250 years, well into the Industrial Revolution. Only when synthetic alum was invented, along with dyes containing their own fixatives, did the last alum works on the Yorkshire coast close, in 1871.

TURNING LEDs BLUE AND WHITE

Light-emitting diodes, or LEDs, are a low-energy light source that also offers smaller, longer-lasting and more robust solutions than traditional bulbs. But until the last few years, the range of colours was limited to red and green – and thus no one could make a practical white light for everyday use.

Researchers knew that they should be able to produce blue light by using a mixture of gallium nitride and indium. But building an LED with the right layers of semiconducting materials was proving difficult. Two pieces of research provided solutions – both in Japan.

GALLIUM AND INDIUM

One technique involved mixing gallium nitride and indium nitride. Today, the InGaN material lights up our blue and green LEDs. Another method was to support gallium nitride on a sapphire wafer – although now it can be done on a cheaper silicon wafer instead.

These steps led to the next breakthrough – being able to coat blue LEDs, usually made with InGaN, with phosphors to form white light.

In recognition of their "invention of efficient blue light-emitting diodes which has enabled bright and energy-saving white light sources", Isamu Akasaki, Hiroshi Amano and Shuji Nakamura were presented with the Nobel Prize for Physics for 2014.

▲ *Isamu Akasaki, Hiroshi Amano and Shuji Nakamura.*

Fast Facts

GALLIUM [Ga]
ATOMIC NUMBER 31

Character: The silvery soft metal called gallium will easily melt in your hand. It has the largest liquid range of any element so far discovered, and does not boil until it reaches 2,400°C (4,352°F).

Discovery: Chemist Paul-Émile Lecoq de Boisbaudran discovered gallium in Paris in 1875. Periodic table pioneer Dmitri Mendeleev had left a gap in his masterwork for gallium to fill, predicting that there should be an element with particular properties that fitted below aluminium.

Name: *Gallia*, the Latin word for France, is said to give gallium its name, although another possibility is that, since *gallus* is Latin for rooster, it is possible Boisbaudran named the element as a pun on the Lecoq part of his name.

World sources: China, Germany, Kazakhstan and Ukraine are among the world's leading producers.

Gallium, indium and tin form an alloy called galinstan, which is non-toxic and can be used in thermometers. ▶

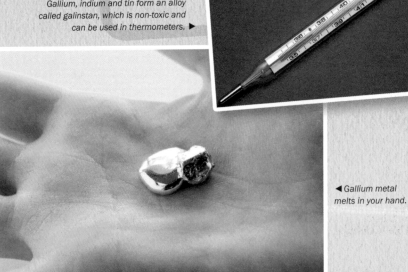

◀ *Gallium metal melts in your hand.*

Fast Facts
INDIUM [In]
ATOMIC NUMBER 49

Character: Stable, soft and silvery, indium is one of the least abundant substances on Earth.

Discovery: German physicist Ferdinand Reich was on the hunt for the recently discovered thallium when he first found indium, in 1863. An unexpected spectral line appeared in a yellow solid he had made from zincblende, and as he was colour blind, he asked a chemist colleague Hieronymous Richter to help with identification. The intense violet line was evidence of an entirely new element.

Name: *Indicum* is Latin for violet and gave the new element its name, indium.

World sources: Indium for industrial applications comes mainly from Canada where it is a by-product of refining ores of zinc, tin and lead.

▲ Indium tin oxide is both electrically conductive and transparent, so makes a useful touch-sensitive coating for screens.

Fast Facts
THALLIUM [Tl]
ATOMIC NUMBER 81

Character: Thallium is another soft, silvery metal and is kept in sealed containers to stop it reacting with the air. It is highly toxic.

Discovery: English physicist William Crookes announced he had discovered thallium in 1861 after he observed a new green spectral line contaminating some sulfuric acid. But Claude-Auguste Lamy in France also found the metal, and extracted enough for the French Academy of Sciences to credit him with the discovery in 1862.

Name: *Thallos* means green shoot in Greek, recalling the bright green of thallium's spectral line.

World sources: Minerals containing thallium are rare, and its production is usually a by-product in smelting copper, zinc and lead.

▲ Thallium takes its name from the Greek word for green shoot.

William Crookes discovered thallium in 1861 and then spent the next ten years working on an accurate measurement of the element's atomic weight.

Fast Facts
NIHONIUM [Nh]
ATOMIC NUMBER 113

Character: Synthetically produced and extremely radioactive, the most stable kind of this super-heavy element has a half-life of only 20 seconds. Its chemical properties are not yet known.

Discovery: A team of scientists at the RIKEN research institute in Japan created element 113 experimentally in 2012. Their finding was confirmed on New Year's Eve in 2015, and they became the first Asian scientific research group to find and name a new element.

Name: The name nihonium comes from *Nihon*, the Japanese name for Japan.

World sources: Groups at RIKEN and also at the Joint Institute for Nuclear Research (JINR) in Dubna, Russia, had tried to create element 113. The Dubna group reported a possible sighting as an alpha decay product of element 115, moscovium. RIKEN's experiment began with a target of bismuth-209, bombarded with zinc-70 nuclei.

◄ The Japanese team who discovered nihonium were the first group in Asia to find and name a new element.

▲ Toxic thallium is colourless, odourless and tasteless, and a favourite of crime and mystery writers.

PERIODIC PIONEERS
KOSUKE MORITA

"There is no short cut to completing any subject." This was a piece of wisdom that pioneer Kosuke Morita shared in an interview for *Asian Scientist* magazine in January 2016. Only weeks before, he had received confirmation from international officials that his group had been recognized as the discoverers of super-heavy element 113, now named nihonium.

Physicist Kosuke Morita. ▶

The road to reach this achievement had been a long one. In Morita's work at RIKEN Nishina Center for Accelerator-based Science, he had seen glimpses of element 113 in experiments during 2004 and 2005. The team had bombarded a target of bismuth-209 with accelerated nuclei of zinc-70 and detected a single isotope, then called ununtrium-278 – an atom of element 113.

However, it was extremely difficult to confirm the finding of this incredibly short-lived element. And pressure was mounting via researchers at the Joint Institute for Nuclear Research in Dubna, Russia, along with Lawrence Livermore National Laboratory in California, who were also on the trail of 113. The Russia-US team has since proven

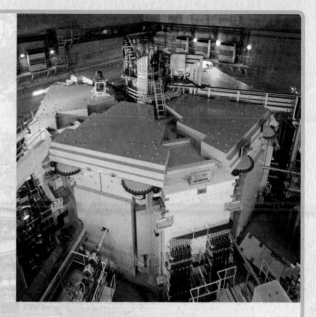

RIKEN's Cyclotron Laboratory. ▶

successful in claiming the discovery of elements 115, 117 and 118.

ELEMENT 113

But finally, in 2012, Morita's team found what they were looking for – proof of all the stages in the radioactive decay chain making the products they saw in their detector. Their discovery that dubnium-262 decays into lawrencium-258 (element 103) and finally into mendelevium-254 (element 101), provided unambiguous proof of element 113 as the origin of the chain.

Kosuke Morita hopes that he will now be able to discover even heavier elements, with atomic numbers of 119 and 120. But he is happy to be part of the first group from Asia to discover and name a new element – and is not someone to pretend that he has always found science straightforward.

Morita loved secondary school physics, and took the subject as an undergraduate. He left Kyushu University in 1984 after completing a doctoral programme in the Graduate School of Science but without having completed his thesis. Nonetheless, he became a research scientist at RIKEN's Cyclotron Laboratory.

But his tenacity has paid off. During the period between first seeing element 113 and securing its discovery, he must have held fast to another wise saying shared in January 2016: "Be an optimist, and think carefully."

6	**C**	Carbon
14	**Si**	Silicon
32	**Ge**	Germanium
50	**Sn**	Tin
82	**Pb**	Lead
114	**Fl**	Flerovium

CHAPTER 8

THE CARBON GROUP

Who really deserved the prize for decoding the molecule of life?

Is glass a liquid or a solid?

Which soft metal provided waterproof flooring for the hanging gardens of Babylon?

THE CARBON GROUP

The fourth vertical group in the periodic table contains carbon (C), silicon (Si), germanium (Ge), tin (Sn), lead (Pb) and flerovium (Fl). It goes by the name of the carbon group, simply because that element appears at the top. However, the endless variety in the chemistry of carbon is so important to life on Earth that this substance merits first place in any listing. From the multitude of personalities pure carbon can take up, to the organic chemicals in which carbon is the universal theme, it is worth taking a long look at carbon's many facets.

That is not to say that the other substances in the carbon group do not pull considerable weight in everyday life. All the elements in the group have

▲ Fossil fuels formed from the remains of ancient carbon-based organisms, through millions of years of intense heat and pressure.

four electrons in their outermost valence shell, and they put this to use in different ways. The semi-metals such as silicon and germanium have a vital role in electronics, because their four-way bonding allows them to form crystal lattices that can be treated so that they conduct electricity.

As you descend the group, there is a marked change from carbon as a non-metal, through silicon and germanium as semi-metals, to tin and lead which are both metals. Flerovium is synthetic and radioactive, and no one is sure exactly what its chemical properties are.

Allotropes of carbon include diamond, in which each carbon atom is covalently bonded to four other carbon atoms in a tetrahedral lattice forming a giant molecular structure. There is also graphite, with carbon atoms bonded in sheets of a hexagonal lattice, and fullerenes, in which carbon atoms bond together in spherical or tube-shaped formations. Graphene, the new carbon on the block, is like graphite in its hexagonal lattice, but consists of a single sheet.

▼ *Allotropes of carbon: diamond, anthracite, graphite and coal dust.*

▲ *Konstantin Novoselov and Andre Geim won the 2010 Nobel Prize for Physics for their discovery of graphene, a new form of carbon.*

ALLOTROPES

Carbon produces some of the most useful allotropes in chemistry. From the Greek word *allos* meaning other, allotropy is the ability of certain elements to exist in more than one physical form while still in the same solid, liquid or gas state. In an allotrope, atoms of the same substance are bonded together in different arrangements, often with distinct properties.

Elements more likely to have allotropes are those capable of existing in a variety of crystal arrangements. Carbon atoms are good at this, as well as at bonding into a chain, a property called catenation.

201

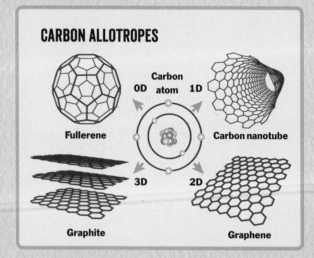

CARBON ALLOTROPES

Fullerene

0D

Carbon atom

1D

Carbon nanotube

3D

2D

Graphite

Graphene

TIN ALLOTROPES

Tin, another element in the carbon group, has two main allotropes, but these perform very differently depending on temperature. Above 13.2°C (55.76°F), tin is a silvery, malleable metal that can be drawn into a wire. This allotrope is called white tin or beta-tin. Drop below the magic temperature, however, and tin objects will gradually decompose into a dense grey powder, a process called tin pest. The element has changed into a different allotrope called grey or alpha-tin, which is brittle, non-metallic and actually has a diamond crystal structure with a higher volume than the beta-tin.

Allotropes of the same elements can have radically different properties. In the case of carbon, diamond is very hard, usually transparent, and does not conduct electricity because the electrons are not free to move but are held tightly between the atoms. Graphite, on the other hand, is flaky, soft and slippery in composition. In its sheet-like structure the layers are only weakly bonded and can easily slide over one another, making graphite the perfect solid lubricant. Since each carbon atom is covalently bonded to three other carbon atoms instead of four, as in diamond, graphite conducts electricity, because the free electrons can move throughout the sheets.

◄ *Carbon dating measures carbon-14 and its decay products in an organic sample to determine its age.*

◀ *Lightweight but strong and stiff, carbon fibre composites are key to building Boeing's 787 Dreamliner and the Airbus A350.*

Fast Facts
CARBON ⒞
ATOMIC NUMBER 6

Character: Carbon ranges in personality between diamond – hard, transparent and inert – and charcoal – brittle, black and combustible. Carbon-14 is a radioactive carbon isotope, produced when cosmic rays bombard nitrogen high in the atmosphere, and is the basis of carbon dating. Since it is taken up by organisms and plants only during life, a measurement of carbon-14's decay products can reveal how much time has passed since death. This finding won Willard Libby the Nobel Prize for Chemistry in 1960.

Discovery: All of life depends on carbon. Carbon found in nature occurs as diamond, graphite, soot, and as anthracite – a type of coal. By heating wood with a limited oxygen supply, early societies could also make charcoal. With the beginnings of chemistry in the 1700s, experimenters started to realize these were all the same element.

Name: The word carbon comes from the Latin word for charcoal, *carbo*.

World sources: In the form of coal, oil and gas, and in minerals like limestone and marble, which are both calcium carbonate, miners extract more carbon from the Earth than any other element.

COLLABORATIVE CARBON

When it comes to bonding, carbon is an over-achiever.
Like all the elements in the group, carbon has four
electrons in its outer shell. However, carbon can make
single, double and triple bonds, as well as arrangements
in which electrons are free to move around. It bonds happily
to itself, in chains that can in theory be infinitely long. And along with key
collaborators, hydrogen, nitrogen, oxygen, phosphorus and sulfur, carbon
makes a colossal range of organic compounds with a variety of uses and
functions. It constitutes and supports all life on Earth.

EARLY ORGANIC CHEMISTRY

In the early 1800s, chemistry was in a
confused state. Swedish chemist Jacob
Berzelius, who had done so much to
measure the atomic weights of the
elements, had confirmed the important
finding that chemicals combine in specific
proportions. But the same did not seem
to apply to the substances found in living
things. All he could say of these was that
they contained the same four elements –

carbon, oxygen, hydrogen and nitrogen
– in varying amounts. Importantly,
he and other leading chemists did not
believe these living substances, which
he called organic, were possible to make.

But over the years that followed,
the notion that living substances must
contain a vital, unknowable spark was
snuffed out. First, one of Berzelius's
former students, a young German called
Friedrich Wöhler, accidently synthesized

▼ Life on Earth relies on carbon.

▲ Urea is now used to make nitrogen-rich fertilizers.

urea in 1828 from non-organic ingredients. This organic compound found in mammalian urine was thus possible to make "without needing a kidney", as Wöhler described his finding, in a letter to Berzelius.

Then in 1845, German chemist Hermann Kolbe managed to make another organic substance, acetic acid. As the idea that organic chemicals required a magic life-giving ingredient faded away, a new enthusiasm grew for deliberate synthesis.

In 1856, English chemist William Henry Perkin was trying to produce the anti-malarial drug quinine, by distilling coal tar and its products. Instead, he made the first synthetic organic chemical dye. Perkin was only 18 years old, but he patented the rich, purple dye called mauveine, and opened a factory to mass-produce it.

Chemists were also making progress with understanding the structure of organic molecules. In the late 1850s, researchers had concluded that each carbon atom could make four bonds, and link together into hydrocarbon chains and lattices. But many mysteries remained. People knew, for example, that the organic substance called benzene contained six carbon atoms and six hydrogen atoms, but how it fitted together as a molecule was still a puzzle.

▲ William Henry Perkin made the first synthetic organic dye, mauveine.

DOUBLE AND TRIPLE BONDS

In a single covalent bond, the atoms share a pair of electrons. However, two carbon atoms can bond together using a **double bond** in which they share four electrons instead of the usual two. While a carbon–carbon double bond is the most common, many elements can bond in this way – carbon dioxide consists of a carbon atom covalently double bonded to two oxygen atoms.

Two carbon atoms can also form a **triple bond**, sharing six electrons. In all cases, the atoms are seeking to create a full outer shell of electrons. Carbon monoxide has a triple bond, and certain other elements can also form triple bonds. Two nitrogen atoms form a stable molecule that makes up the majority of the air we breathe.

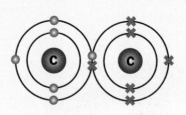

Carbon atoms can bond together with a single covalent bond.

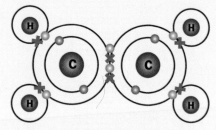

They can also form double bonds, sharing four electrons, as in the hydrocarbon ethene.

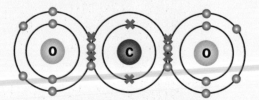

Double bonds can form between carbon and oxygen, in carbon dioxide.

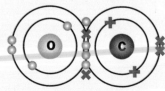

Carbon monoxide contains a carbon and an oxygen atom connected with a triple bond.

AROMATIC COMPOUNDS

Since 1855, benzene and its compounds had been known as aromatics after August Wilhelm Hofmann (of whom William Henry Perkin was a student) recognized many of them were derived from substances with a pleasant smell. By the early 1860s, scientists thought perhaps benzene contained several double bonds, or that it formed into rings. But it was not until 1865 that history books record Friedrich August Kekulé's suggestion that it was a ring of six carbon atoms with alternating single and double bonds between them.

According to an account Kekulé gave later in life, he conceived the idea of the ring formation of benzene when he nodded off to sleep while writing a student textbook. As chains of carbon atoms danced around in his dreams, he saw one curl into a ring like a snake and bite its own tail. While other people may also have proposed the ring structure, Kekulé's explanation had a particular charm.

▲ *German chemist August Wilhelm Hofmann.*

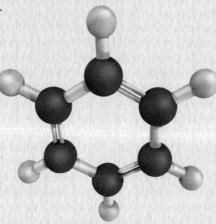

◄ *Kekulé said he conceived his ring structure for benzene when he dreamed of a snake biting its own tail.*

207

The purple dye mauveine, it would turn out, was a mixture of aromatic compounds – and the first synthesized medicine was also based around an aromatic ring. In 1899 came the first success of the push to synthesize new drugs using the flourishing new science of organic chemistry. Scientists working at the Bayer Company in Germany launched the now-ubiquitous drug aspirin, acetylsalicylic acid. It relieved pain without the gastric side-effects of salicylic acid, an extract of willow.

Aromatics had brought the sweet smell of success to the earliest years of both the dye industry and the pharmaceutical industry, as chemists began to work on new organic products.

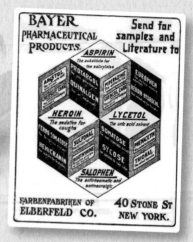

▲ Bayer was the first company to market painkiller aspirin.

▼ Medicines made from willow and other plants rich in salicylates have a long history.

▲ Irish physicist Kathleen Lonsdale.

PROVING BENZENE'S STRUCTURE

Chemists accepted the aromatic ring structure of benzene. But still nobody knew quite how the carbon and hydrogen atoms sat within the circular structure. Was it a flat ring in a single plane? Or did the six carbon atoms form a sort of zigzag arrangement?

The answer came – but not until 60 years later, with the invention of X-ray crystallography. This technique, developed by father-and-son William Henry Bragg and William Lawrence Bragg, was the first method that allowed chemists to work out the shape of organic molecules within crystals.

One of Bragg senior's students was an Irish-born physicist called Kathleen Lonsdale. She had moved to England at the age of five, and learned mathematics and science in a boys' school because her own did not offer lessons at an advanced enough level.

After gaining a degree, Lonsdale worked with Bragg at University College London, where she used X-ray diffraction to investigate the structure of a benzene-based substance called hexamethylbenzene. The advantage of this molecule was that it is solid at room temperature, allowing Lonsdale to fire X-rays at it and look at the resulting patterns.

Her work showed, in 1928, that the benzene ring was definitely flat, and that the carbon–carbon bonds in the ring were all the same length – answering one of the most tantalizing questions in organic chemistry at that time.

SYNTHESIS

The finding was significant, not only for aromatic molecules themselves but because it helped to turn the tide toward confirming the synthesis of chemicals via their structures. Lonsdale went on to study the synthesis of diamond, and this gave her another claim to fame. Her name is remembered in lonsdaleite, a kind of diamond only found in meteorites.

▶ Lonsdaleite forms when meteorites containing graphite hit the Earth, transforming the graphite through heat and stress.

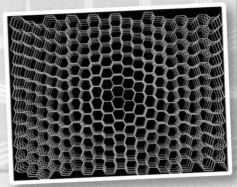

OIL TO PLASTIC

Along with the aromatic molecules, there was a second huge family of hydrocarbons recognized by 19th-century chemists – the aliphatic compounds. These were named after the Greek word for fat, because they could be derived from fats and oils.

In the 1850s, entrepreneurs built the earliest industrial crude oil production plants in pursuit of oil for lamp-lighting. James Young, a chemist from Scotland, refined crude oil and oil shale, patented the process, and built a commercial oil refinery in 1851.

A Polish pharmacist called Jan Józef Ignacy Łukasiewicz was also keen to put petroleum and its products to good use. Along with a colleague, Jan Zeh, Łukasiewicz built an effective kerosene lamp after distilling petroleum to create a steady-burning fuel. They patented their invention, which by 1853 was in use in hospitals. With a mine-owner, Łukasiewicz then began operating an oil field near the village of Bóbrka, along with flourishing refineries.

Drilling for oil grew into an industry in the USA, Canada, Russia and parts of Europe, and in 1913 a chemist called William Merriam Burton invented the first thermal process for cracking. This was the way to break up long chains of hydrocarbons in oil into shorter ones, providing a source of petrol, naptha and diesel.

▼ *Oil well equipment at the Ignacy Łukasiewicz Museum of Petroleum Industry in Bóbrka, Poland.*

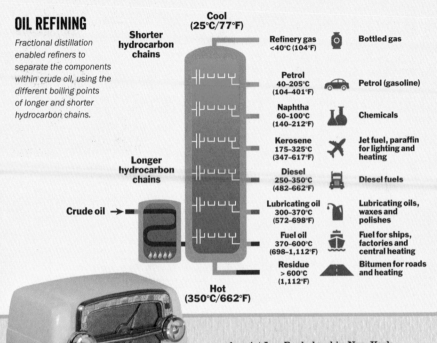

OIL REFINING

Fractional distillation enabled refiners to separate the components within crude oil, using the different boiling points of longer and shorter hydrocarbon chains.

Cool (25°C/77°F)

Shorter hydrocarbon chains

Longer hydrocarbon chains

Crude oil →

Hot (350°C/662°F)

Refinery gas <40°C (104°F)		Bottled gas
Petrol 40–205°C (104–401°F)		Petrol (gasoline)
Naphtha 60–100°C (140–212°F)		Chemicals
Kerosene 175–325°C (347–617°F)		Jet fuel, paraffin for lighting and heating
Diesel 250–350°C (482–662°F)		Diesel fuels
Lubricating oil 300–370°C (572–698°F)		Lubricating oils, waxes and polishes
Fuel oil 370–600°C (698–1,112°F)		Fuel for ships, factories and central heating
Residue >600°C (1,112°F)		Bitumen for roads and heating

▲ *Popular Bakelite gadgets helped bring in the age of plastics.*

Shorter chains of hydrocarbons were also the ingredients of a new revolution in materials – plastics. These were long chain-like molecules – synthetic polymers – made of repeated units. Bakelite was an early plastic, invented in 1907 by Belgian chemist Leo Baekeland in New York. It was heat-resistant, an electrical insulator and relatively cheap compared with the natural polymers then available.

Bakelite was so successful that, after the First World War, the age of plastics really took off. Technology advanced so that chemists could mass-produce polymers from repeated carbon-based units with different properties. These formed the familiar plastics we know today: PVC in the 1920s, polythene in the 1930s and polypropylene in the 1950s.

STEPHANIE KWOLEK AND KEVLAR

American chemist Stephanie Kwolek once thought she might become a doctor. But the material she invented was five times stronger than steel, and a vital component of life-saving bulletproof vests and body armour. Its trade name was Kevlar.

▲ *Chemist Stephanie Kwolek.*

Kwolek joined the US chemical company DuPont in 1946 and began investigating petroleum-based fibres for their strength and rigidity. Her specialism was in synthesizing the intermediate substances, called aromatic amides, which could then be dissolved in solvents and spun into fibres.

Most of the solutions Kwolek made were thick and clear, but one day in 1965, she created a runnier, cloudy substance. She experimentally spun the fibre, and it turned out to be stiffer than any previously created, with very high strength when stretched.

What was the explanation? Kwolek discovered it was because, under certain conditions, the polyamides she was working with formed liquid crystals while in solution. As the fibres were drawn out during spinning, large numbers of these organic chains lined up in parallel. Strong intermolecular forces called hydrogen bonds between the chains accounted for the fibre's incredible strength.

Kwolek led polymer research in DuPont's Pioneering Lab until she retired in 1986. She received the Perkin Medal in 1997, an award given by the Society of Chemical Industry America in recognition of outstanding work in applied chemistry, and first awarded to William Henry Perkin, discoverer of the first synthetic organic dye.

◄ *One of Kevlar's applications is in bulletproof vests.*

MOLECULES OF LIFE

▼ Crowfoot Hodgkin discovered the structure of penicillin using X-ray crystallography.

Research into life-saving organic pharmaceuticals gathered pace. In 1909, a team led by Paul Ehrlich, a German chemist, discovered that an organic chemical containing arsenic and numerous carbon rings was effective against the rampant killer disease syphilis. The drug, arsphenamine, worked by killing the bacteria responsible for the infection, without killing the human patient, and opened up the new field of chemotherapy.

Structures of natural organic materials were also being illuminated by the technique of X-ray crystallography. In 1928, a British undergraduate chemist at Oxford University called Dorothy Crowfoot (later Crowfoot Hodgkin) made

▲ Nobel Prize-winning chemist Dorothy Crowfoot Hodgkin.

some of the first X-ray studies of organic molecular structure. This set the scene for a career in which she unravelled the structure of penicillin using X-ray diffraction in 1945, and then vitamin B12 in the 1950s. She received the Nobel Prize for Chemistry for this work in 1964, and went on to decipher the structure of insulin. During Crowfoot Hodgkin's Ph.D studies at Cambridge University in the early 1930s, her advisor had been physicist John Bernal. Three of his other students, at Cambridge and later at Birkbeck College London, also made remarkable progress in deciphering the structures of key organic molecules. Max Perutz, a British biochemist born in Austria, shared in the 1962 Nobel Prize for Chemistry for working out the structure of haemoglobin, which carries oxygen around the body. Lithuanian-born

◀ Maurice Wilkins.

◀ Rosalind Franklin.

▲ Linus Pauling.

British chemist Aaron Klug developed the new technique of crystallographic electron microscopy to study proteins and viruses, for which he received the Nobel Prize for Chemistry in 1982.

ROSALIND FRANKLIN

British scientist Rosalind Franklin was another of the high-flying students who worked with John Bernal. She joined his team at Birkbeck in 1953 to study the structure of viruses, and built a thriving and productive research programme before her untimely death from cancer at only 37, in 1958.

But it was the work she had done during 1951 and 1952, at King's College London, for which she is really remembered. In a team that also included chemist Maurice Wilkins, Franklin worked to understand which organic molecules might be the secret of heredity between generations. Scientists at the Rockefeller Institute for Medical Research in New York had proposed that DNA was the vital substance, while other researchers thought proteins were more likely. Franklin brought to bear her expertise in X-ray diffraction in trying to understand the structure of DNA – deoxyribonucleic acid.

Wilkins, Franklin and the group at King's were far from the only people interested in DNA. Back in the USA, Linus Pauling had worked out, in spring 1951, that proteins folded up by forming helices and sheets. The finding later won him the Nobel Prize in Chemistry – but now he moved on to the structure of DNA. Meanwhile, at Cambridge University, James Watson and Francis Crick were trying to build a physical model of DNA, using clues to the molecule's characteristics that were emerging from King's College and other London University teams. In November 1951, Crick and Watson produced a triple helix model they thought might work. It had a chain of phosphates as a kind of backbone at its centre, and when she saw the model, Franklin could see straight away that this configuration would not hold together.

DNA'S COMPONENTS

The components of DNA were well known to all the competitors in the race. As early as 1878, German biochemist Albrecht Kossel had isolated nucleic acid, and later found five bases within it – adenine, cytosine, guanine, thymine and uracil. For this work he received the 1910 Nobel Prize in Physiology or Medicine. In 1919, American biochemist Phoebus Levene worked out that there were phosphate, sugar and base units in DNA, which he called nucleotides. He suggested DNA

THE FIVE BASES IN DNA AND RNA

Purines are fused-ring organic molecules, while pyrimidines are simple ring organic molecules. Within the DNA molecule, adenine pairs with thymine and guanine pairs with cytosine. Uracil replaces thymine in the molecule RNA, which creates proteins from DNA.

PURINES

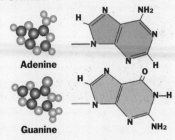

Adenine

Guanine

PYRIMIDINES

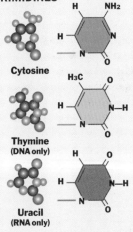

Cytosine

Thymine
(DNA only)

Uracil
(RNA only)

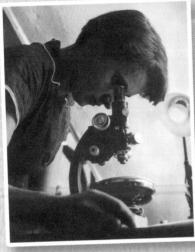

◄ Rosalind Franklin and her student Raymond Gosling carried out vital experimental work on the structure of DNA.

might be a string of these nucleotides, linked together with the phosphate groups. In Russia in 1927, biologist Nikolai Koltsov had even proposed that traits would pass from one generation to the next via a giant hereditary molecule containing two mirror strands that would replicate – although his ideas probably did not reach those working on DNA in the West.

In the lab at King's, Franklin and her Ph.D student Raymond Gosling continued to study DNA. They knew the density of the molecule and that the phosphates, which give DNA its structure, would be on the outside. However, the X-ray results were not clear-cut. In one form of DNA, called wet, the molecule appeared to show evidence of being a helix. One of

the images captured by Franklin in May 1952 – known as photograph 51 – showed a beautifully clear diffraction pattern of the wet form of the molecule that corresponded with a helical structure. In DNA's dry form, however, the format was harder to determine.

Although work was proceeding well, relations were strained between Franklin and Maurice Wilkins. Miscommunications meant that she often felt her expertise was underestimated. She arranged to move to work at Birkbeck College and wrote up her experiments in preparation to leave King's – and DNA research – behind. In doing so, she re-examined her data on both wet and dry DNA and concluded they both showed a two-chain helix, although she had not worked out how the bases inside were arranged.

DECODING PHOTOGRAPH 51

Rosalind Franklin's image, called photograph 51, helped make the leap between knowing the constituents of DNA, to realizing its structure. To capture the image, Franklin and her student Raymond Gosling mounted a tiny sample of DNA inside a camera, where they bombarded it with X-rays for more than 60 hours. The beam of X-rays scattered and produced the photograph.

The darker patches are areas of the film that have been hit multiple times by diffracted X-rays, and these correspond to regular, repeating features within the molecule. Although detailed mathematics is needed to interpret the image, the cross-shape indicates a helix, with each arm representing the zig or zag of the spiral as seen from the side. The dots along the arms of the cross show that there are ten bases in each turn of the helix. By its symmetry, the image also shows that the DNA crystal has the same form when turned upside down. An intriguing clue to DNA's helical structure is the missing fourth blob as counted out from the centre along the arms of the crosses. This is because the DNA helices are slightly offset with respect to their turn around a central axis, giving the major and minor grooves shown on the image.

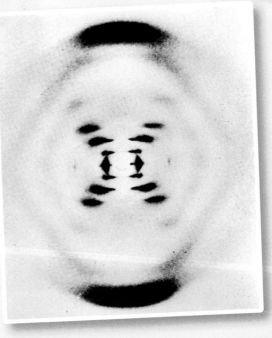

▼ Photograph 51 was the key piece of evidence for DNA's two-chain helix structure.

In February 1953, the race to unravel DNA intensified. Pauling produced another triple-stranded helical model, but Watson and Crick at Cambridge knew that this was incorrect. They had now had sight of some of Franklin's unpublished data – along with photograph 51, which Wilkins had shown them after Raymond Gosling had passed it to him in the course of his Ph.D work. Crick and Watson also knew, via a meeting with Austro-Hungarian biochemist Erwin Chargaff, that the quantity of adenine in DNA matched that of thymine, and the amount of cytosine matched that of guanine.

▲ *James Watson and Francis Crick with their model of part of a DNA molecule.*

▲ *Erwin Chargaff, Austro-Hungarian biochemist.*

As Crick and Watson experimented with different models, they realized that if adenine bonded with thymine, it took up the same space as cytosine bonded with guanine. Thus, these pairs of bases could run between two helical strands of structural sugar-phosphate chains. This idea fitted with all the evidence: DNA was a two-stranded helix with base pairs forming the steps of a spiral staircase-like structure. Crick and Watson announced their finding in *Nature* on 25 April 1953. The journal also included an accompanying article by Franklin and Gosling on their work

THE DOUBLE HELIX

Life as we know it springs from the double helix of DNA, which contains hydrogen, oxygen, nitrogen and phosphorus along with carbon. The backbone of DNA is made up of phosphates – a phosphorus atom bonded to four oxygen atoms – alternating with five-carbon sugar molecules. Attached to the backbone are the bases. Via intermolecular bonds called hydrogen bonds, adenine bonds with thymine, and guanine bonds with cytosine, holding the molecule together.

Minor groove

Major groove

○ Hydrogen
◎ Oxygen
○ Nitrogen
● Carbon
○ Phosphorus

T   A

C   G

Pyrimidines   Purines

– although it was not clear in Crick and Watson's paper how much Franklin's X-ray diffraction photos, and other data, had helped them in their discovery.

In 1962, Francis Crick, James Watson and also Maurice Wilkins, who had continued to work on the structure of DNA, received the Nobel Prize for Physiology or Medicine. Rosalind Franklin's role in the discovery has only become clearer in recent years.

▲ In April 1953, Nature published Crick and Watson's paper on DNA structure, as well as an article by Rosalind Franklin about her X-ray diffraction photos.

SILICON AND THE SEMICONDUCTORS

Just as organic life relies on a carbon-based ecology, you could argue that modern technology has developed its own silicon-based ecosystem, with its headquarters in Silicon Valley. Everywhere you look, silicon is colonizing the world, from the microchip in your smartphone, laptop, washing machine or car, to the wafer in a solar cell on your roof.

Silicon's success in the electronic landscape is because of its properties as a semiconductor. The electrons are usually trapped within silicon so it does not conduct electricity – just like an insulator. However, as you warm silicon, some of the electrons are promoted into a conduction band where they can move freely and conduct electricity.

At the same time, the holes left behind in the silicon's crystal lattice can move too. By adding other elements to the silicon – a process known as doping – you can modify its electronic properties.

In the periodic table, there is a huge constituency of metals on the left and a considerably smaller zone of non-metals on the right. In the hinterland between,

▼ Silicon's properties as a semiconductor make solar panels function.

▲ In space, satellites can deploy solar panels, making use of the Sun's energy.

we find a zigzag-shaped stripe consisting of metalloids, including boron in the boron group, silicon and germanium in the carbon group, arsenic and antimony in the nitrogen group, and tellurium and polonium in the oxygen group. These substances are also semiconductors. Silicon, along with germanium, plus compounds gallium arsenide, lead sulfide and indium phosphide, are all semiconducting materials that have hugely varied applications in electronic devices.

THE BAND GAP

Normally, electrons occupy the valence band. In a semiconductor, if you provide the right amount of heat, light or voltage, they will jump to the conduction band.

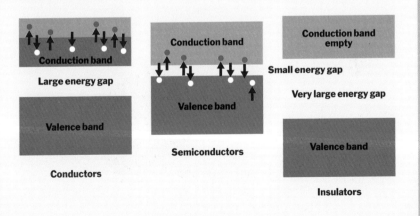

DOPING SEMICONDUCTORS

Even a small addition of atoms to the regular crystal structure of silicon or germanium makes a big difference to its properties. If you add atoms of an element with five outer electrons – antimony, arsenic or phosphorus, for example – you end up with additional free electrons in the system, and create an N-type semiconductor (after negative).

If you add atoms of an element with only three outer electrons – boron, aluminium or gallium – you create additional holes in the system, and create a P-type semiconductor (after positive).

Junctions between P and N semi-conductors are the basic building blocks of most electronic devices in the form of diodes and transistors.

SOLAR CELLS AND P-N JUNCTIONS

A silicon solar cell contains an ultra-thin layer of phosphorus-doped (N-type) silicon, with free electrons inside, plus a thicker layer of boron-doped (P-type) silicon. Where these two layers are in contact, an electrical field is created at a P-N junction. When sunlight falls on the cell, it gives extra energy to millions of free electrons, which move across the P-N junction guided by the electrical field to provide a useful flow of current.

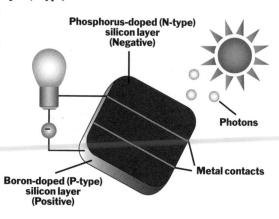

Phosphorus-doped (N-type) silicon layer (Negative)

Photons

Metal contacts

Boron-doped (P-type) silicon layer (Positive)

▲ By melting silica, along with other ingredients, people have been making glass for 3,000 years.

▲ Silicones, used for dental imprints and for waterproof sealants, are long molecular chains made of silicon bonded to oxygen in a repeating pattern.

Fast Facts

SILICON Si

ATOMIC NUMBER 14

Character: Pure silicon crystals have a blue-grey sheen, and are semiconductors – being neither fully conducting like metal, or fully insulating.

Discovery: Silica, also known as silicon dioxide or quartz, was one of the first minerals used by humans in the form of flints. Quartz crystals are used in clocks and other electrical devices because of the piezoelectric effect – they produce a tiny electrical current when compressed.

Name: The word silicon comes from the Latin *silex* or *silicis,* meaning flint.

World sources: Silicon is second only to oxygen in terms of its abundance in the Earth's crust. As well as silicate sand and stone that occurs all over the world, there are some precious versions of silica – opals, rock crystal, rhinestone and the purple stone amethyst, which contains traces of iron.

IS GLASS A LIQUID OR A SOLID?

Ancient church windows make the most of the jewel-like transparency of stained glass. But for many years a myth persisted that in the oldest panes you would find a greater thickness at the bottom because the glass had flowed downward.

▼ *Traditional glassblowing can produce glass thicker on one side.*

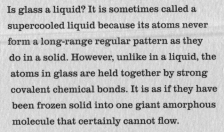

Is glass a liquid? It is sometimes called a supercooled liquid because its atoms never form a long-range regular pattern as they do in a solid. However, unlike in a liquid, the atoms in glass are held together by strong covalent chemical bonds. It is as if they have been frozen solid into one giant amorphous molecule that certainly cannot flow.

When glassmakers blew and spun the glass, it led to pieces of glass thicker at one end – and on the whole they chose to place them with the heavier side downward – hence the myth.

Fast Facts
GERMANIUM [Ge]
ATOMIC NUMBER 32

Character: Pure germanium is silver-white and brittle, and is a stable metalloid. Its semi-conducting properties made it the focus of a key experiment in December 1947, carried out by John Bardeen, William Shockley and Walter Brattain at Bell Laboratories in the USA. In their search for a solid-state replacement for unreliable, bulky vacuum tubes, they discovered that germanium amplified a small signal into a larger one – the transistor effect, for which they received the Nobel Prize for Physics in 1956.

Discovery: Dmitri Mendeleev predicted the existence of an element between silicon and tin and, 14 years later, in 1885, an unusual ore turned up in a silver mine in Freiburg, Germany. At the nearby University of Mining and Technology, Clemens Winkler analyzed the ore and found it contained 7 per cent of a substance he could not identify. By the following year, he had realized that by its properties it must be the missing carbon group element.

Name: Germanium takes its name from the Latin word for Germany, *germania*.

World sources: China, Russia and USA all produce germanium as a by-product of zinc ore.

▲ *Nobel Prize-winning inventors of the transistor: John Bardeen, William Shockley and Walter Brattain.*

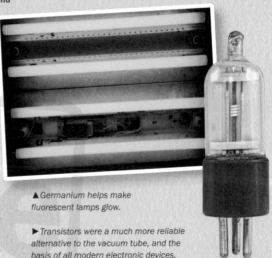

▲ *Germanium helps make fluorescent lamps glow.*

▶ *Transistors were a much more reliable alternative to the vacuum tube, and the basis of all modern electronic devices.*

Fast Facts

TIN Sn

ATOMIC NUMBER 50

Character: Tin is a soft, workable metal which is sensitive to lower temperatures and can turn into powder below 13°C (55.4°F). Food cans were originally made of steel electrolytically coated with tin to avoid corrosion. Even though aluminium and other metals have now taken the place of tin, the name "tin can" has stuck.

Name: The word tin has a Germanic root, and the element's chemical symbol Sn comes from the Latin word *stannum*.

Discovery: Thousands of years ago, when an enterprising early chemist discovered that adding a small amount of tin to copper made a harder, longer-lasting metal, the Bronze Age was born. Trade and agriculture expanded, systems of writing and medicine developed, and people built and lived in cities, beginning in the Ancient Near East in around 3300 BCE.

World sources: A tin belt runs through China, Thailand and Indonesia, where the ore cassiterite provides the main source of tin, and the metal is also mined in South America.

▲ *Before plastics, tin toys were a cheap and brightly coloured childhood treat.*

▶ *Sir John Franklin set off in 1845 to find a sea route linking the Atlantic to the Pacific; all his crew succumbed to malnutrition and disease.*

METAL MISADVENTURES

▲ *Robert Falcon Scott.*

Since the 19th century, the humble tin can has helped explorers make journeys and expeditions, which to that point had been all but impossible. But there is a darker side to this ancient metal.

While tin cans helped Scott of the Antarctic preserve food supplies, for example, he did not know that the metal decays at low temperatures. This degradation, known as tin pest, created tiny holes in tin solder holding together the fuel storage containers he was relying on. The kerosene leaked away, leaving the team at the mercy of the Antarctic blizzards in March 1912.

Until recently, lead solder in food cans also stood accused of contributing to the death of Sir John Franklin and his crew on their 1845 expedition to find the Northwest Passage. However, more recent evidence has absolved the cans – and lead water pipes on the ships – of their guilt in this case.

In 2016, a team at TrichAnalytics in Canada analyzed the fingernails of those who perished, to find that zinc deficiency from malnutrition rather than lead poisoning was to blame. A lack of zinc would suppress the immune system, leaving Franklin's crew at far greater risk of death from tuberculosis or pneumonia.

Fast Facts

LEAD [Pb]
ATOMIC NUMBER 82

Character: Dull and silvery-grey, lead is soft and malleable, with many uses in history. However, it is highly toxic and interferes with the work of enzymes in the body. It also crosses the blood-brain barrier, causing neurological damage.

Name: The word lead has Germanic origins and the chemical symbol Pb is from *plumbum*, its Latin name.

Discovery: Lead beads found in central Anatolia – modern-day Turkey – are at least 9,000 years old, and may be an example of the earliest metal-smelting in the world. Lead is rare in its metallic state and so would have had to be extracted from ore. In the ancient world, lead also featured in ornaments from Egypt, coins in China, and even water-retaining flooring for the Hanging Gardens of Babylon in Assyria, modern-day Iraq. The Romans needed lead for their complex plumbing and water supply systems, as well as for making pewter pots and pans, and in shipbuilding.

World sources: Many countries have lead deposits, mostly in the form of galena, a lead-bearing mineral.

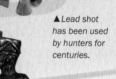

▲ Lead shot has been used by hunters for centuries.

◄ Toy soldiers cast in solid lead were popular in Germany during the 19th and early 20th centuries.

▲ Lead may have given the Hanging Gardens of Babylon a useful waterproof flooring material.

▲ Lead-glazed earthenware is one of the traditional varieties of ceramic.

▼ Merchants could identify and secure goods in transit by adding a stamped seal made of lead.

▲ Queen Elizabeth I wore white lead and vinegar on her face for a pale appearance, but the toxic metal had a terrible effect on her health.

▼ The Romans required lead for their sophisticated plumbing systems and bath houses, as in these archaeological remains in Beirut, Lebanon.

JAGADISH CHANDRA BOSE AND THE LEAD SEMICONDUCTOR

Born in 1895 in the country now called Bangladesh, Jagadish Chandra Bose was a polymath. His interests included electromagnetic waves, plant physiology, and even writing science fiction. And despite the racial discrimination of British Indian rule under which he lived, which often denied him access to laboratory space, Bose had 27 papers published in the journal *Nature* during his long career.

▲ *Polymath Jagadish Chandra Bose was an early pioneer in the development of semiconductors.*

In 1895, Bose made the world's first public demonstration of radio waves, in the town hall in Kolkata. At an impressive distance, he rang an electric bell and set off a gunpowder charge, showing the waves could pass through solid objects and be detected remotely.

Bose also made key early discoveries about semiconductor devices. He found that galena, a natural mineral form of lead sulphide, could selectively conduct electricity in the presence of radio waves, making it a useful detector.

As a habit, Bose resisted patenting his discoveries, which is one reason that his contributions are less well-known than you might expect. But in 1904 he did patent his "detector for electrical disturbances", describing substances

▲ Sitting (from left): astrophysicist Meghnad Saha, Jagadish Chandra Bose, chemist Jnan Chandra Ghosh. Standing (from left): physicist Snehamay Dutta, quantum physicist Satyendra Nath Bose, particle physicist Debendra Mohan Bose, physicist Nikhil Ranjan Sen, chemist Jnanendra Nath Mukherjee, chemist Nagendra Chandra Nag.

that showed either lower or higher resistance with increasing voltage. He called these positive and negative substances – an interesting parallel with the P and N terminology of modern semiconductors. Pioneer of the transistor, Walter Brattain, always credited Bose with having been ahead of the game in semiconductors – a field that only really gained ground in the 1930s.

INDIA POSTAGE

30·11·1858 – 23·11·1937
जगदीश चन्द्र बोस
JAGADISH CHANDRA BOSE

15 नo पo
nP

COLLABORATION TO MAKE SUPER-HEAVY ELEMENTS

During the Second World War, nuclear physicist Georgy Flerov was serving in the Russian air force. He noticed that work on nuclear fission in the scientific literature had ceased to appear in American, British and German publications – and deduced that it had become classified research as the different powers worked toward atomic weapons. As the result of a letter Flerov wrote to Stalin in 1942, a Soviet atomic bomb project began – along with an arms race lasting decades.

Leap forward 40 years, however, and the field of battle had become one of scientifically powered discovery rather than destruction. Flerov's pioneering work in heavy-ion physics meant he was now leading his own laboratory at the Joint Institute for Nuclear Research (JINR). In both Russia and America, scientists were engaged in the race to find super-heavy elements, those with atomic numbers greater than 104,

▲ Georgy Flerov's work led to the discovery of two new elements. On his wall is an image of Igor Vasilyevich Kurchatov, a nuclear physicist who tested the first Soviet atomic bomb in 1949; Kurchatov was nicknamed "the beard" after saying he would not shave until the programme was successful.

▶ The Joint Institute for Nuclear Research (JINR) is in the city of Dubna, Russia.

not found in nature because they decay radioactively within milliseconds.

It was a competitive field – but then, at a conference in 1989, two giants of nuclear physicists met: Flerov of JINR and Ken Hulet from the Lawrence Livermore National Laboratory in California, USA. Despite the prevailing political scene – including the fact that both countries were still testing nuclear weapons – the two scientists agreed to collaborate.

SUPER-HEAVY DISCOVERY

The first visit of US personnel to the lab at JINR took place in 1990, and shortly after that Flerov died. But by 1998 the collaboration yielded discovery of the super-heavy elements 114 and 116.

According to convention, the new elements were named by the scientists who discovered them. In 2012, flerovium and livermorium, from the respective research institutes, were announced by the International Union of Pure and Applied Chemistry. While the elements are not stable enough for applications beyond experimentation, they were testament to an international collaboration built on a firm foundation in an ever-changing world.

Fast Facts
FLEROVIUM [Fl]
ATOMIC NUMBER 114

Character: As it lasts such a short time before decaying, flerovium is primarily just known as one of the super-heavy, radioactive elements. However, there is some evidence that it is unexpectedly volatile for an element in this group, becoming a gas at room temperature, and is the least reactive element in its group.

Name: Flerovium gained its name in 2012, recalling the distinguished Russian physicist Georgy Flerov. The laboratory where the element was discovered is named after him.

Discovery: A team led by Yuri Oganessian and Vladimir Utyonkov in the Flerov Laboratory of Nuclear Reactions of the Joint Institute for Nuclear Research (JINR) in Dubna, Russia, managed to create flerovium in 1998. The experiment to make flerovium lasted 40 days, during which they fired 5 billion billion calcium atoms at a target of plutonium. A single atom of flerovium, decaying after 30.4 seconds, was detected.

World sources: In total, about 90 atoms of flerovium have ever been made, with mass numbers from 284 to 290, according to how many neutrons are in their nuclei.

7	
N	
Nitrogen	

15	
P	
Phosphorus	

33	
As	
Arsenic	

51	
Sb	
Antimony	

83	
Bi	
Bismuth	

115	
Mc	
Moscovium	

CHAPTER 9

THE NITROGEN GROUP

HOW DID NITROGEN FULFIL ALCHEMISTS' DREAMS?

WHICH ELEMENT IN THIS GROUP WAS FIRST FOUND IN BUCKETS OF HUMAN URINE?

WHY WOULD YOU HAVE SMEARED ANTIMONY ON YOUR FACE IF YOU HAD LIVED IN ANCIENT EGYPT?

LIFE AND DEATH IN THE PNICTOGENS

As well as having the straightforward name of the nitrogen group, this collection of elements sometimes goes by the unpronounceable-looking title of pnictogens. Derived from the Ancient Greek *pnigein*, meaning to choke, the name refers to the stifling property of nitrogen.

Early scientists trying to understand gases named oxygen "vital air", carbon dioxide "fixed air", and hydrogen "inflammable air". Nitrogen, meanwhile, was given the name "noxious air" by its discoverer Daniel Rutherford, and "mephitic air", meaning toxic to humans, by the great chemist Antoine Lavoisier.

Within the group are elements that both sustain life and snuff it out. Some substances are in both camps, depending on the application. Nitrogen and phosphorus are vital as components of the nucleic acids DNA and RNA, and the molecule ATP that transports energy in our cells.

▲ *Bismuth and antimony sulfate produce crystals with striking colours and shapes.*

▲ Nitrogen and hydrogen react together to produce ammonia, a base for the production of fertilizers, dyes and explosives.

DEADLY ELEMENTS

However, nitrogen and phosphorus also offer some of the world's deadliest poisons in the form of cyanide and white phosphorus. Arsenic is another nasty chemical element in this group.

Further down the group, antimony and bismuth both played roles in beautifying the Ancient Egyptians as key ingredients in cosmetics. These two elements themselves form strikingly pretty crystals. Finally, we reach moscovium. Is it deadly, life-giving, a blessing or a curse? Nobody yet knows.

SOLIDS AND GASES

With five electrons in their outer shell, the nitrogen group elements can each form three bonds, and all are solids with the exception of gaseous nitrogen.

Why, though, should nitrogen be a gas in everyday life, while the element to its left in the periodic table, carbon, forms a solid at room temperature? In fact, atomic weight bears little relation to whether a substance is a solid, liquid or gas. What makes the difference is the forces between atoms or molecules of the element.

Carbon forms a lattice at room temperature and pressure, with strong forces between the atoms that hold them together. The solid needs to reach a very high melting point to turn into a liquid, and even higher to boil. Nitrogen, on the other hand, forms into two-atom molecules, which do not readily attract one another. They have no reason to stick together as a liquid or solid, so at room temperature the element is a gas.

ALCHEMY'S PROMISE FULFILLED IN NITROGEN

Nitrogen was the first stable element to fulfil the alchemists' dream of transmuting into another substance. In 1917, Ernest Rutherford was working at Manchester University in the UK where he had already discovered that the atom had a tiny, dense nucleus. From earlier investigations into alpha and beta radiation, he knew that elements could spontaneously turn from one to another through radioactive decay – this was the basis of his Nobel Prize in 1908.

But he had also seen that alpha particles produced an unknown radiation when beamed into the air, and wanted to investigate further. Rutherford thought the nitrogen in the air was key to the effect, so he bombarded pure nitrogen with alpha radiation. Two outcomes resulted: the mystery radiation turned out to be hydrogen nuclei, and he found he had managed to turn nitrogen into oxygen – a nuclear transmutation of one stable element into another:

nitrogen-14 + helium-4 → oxygen-17 + hydrogen-1

Now, there is no great urgency to create oxygen through arcane processes, as the early alchemists tried to do with gold. But the experiment did yield a

LIFE-SAVING EXPLOSION

When a car airbag is triggered by a collision, an electrical spark ignites a chemical – often sodium azide – which rapidly fills the airbag with nitrogen gas.

Crash sensor

Airbag

Steering wheel

Inflator

Airbag

Nitrogen gas

▲ *Sperm and embryos are cryopreserved using liquid nitrogen in the process of IVF.*

precious finding. It gave Rutherford the clue that hydrogen nuclei were part of nitrogen nuclei, and might therefore be a component of every nucleus, leading him to postulate the existence of a new elementary particle, the proton.

▼ *Hair dyes sometimes contain ammonia to help the hair absorb colour.*

Fast Facts
NITROGEN [N]
ATOMIC NUMBER 7

Character: Colourless and odourless, nitrogen is a generally nonreactive gas at room temperature. However, nitrogen is vital to life, a crucial ingredient in building proteins. Your body contains 3 per cent nitrogen by mass.

Discovery: While studying for his doctorate in 1772, Scottish chemist Daniel Rutherford proposed that nitrogen was a chemical element, and the main constituent of air. Earlier experiments had shown there was a constituent of air in which fire would not burn and a mouse could not survive – but no one had claimed it as an element.

Name: Nitrogen comes from the Greek words *nitron genes*, meaning that it makes nitre. This was the former name for potassium nitrate, also known as saltpetre. It was a vital ingredient in the manufacture of gunpowder.

World sources: 78 per cent of the air around us, by volume, consists of nitrogen. In the chemical industry every year, nitrogen and hydrogen react together to produce millions of tonnes of ammonia, a base for fertilizers, dyes and explosives.

COMPLETING THE NITROGEN CYCLE

Sergei Winogradsky was 93 years old when he completed a 900-page book – *Soil Microbiology: Problems and Methods, Fifty Years of Investigations*. His long career had taken him from his home town of Kiev to work in Strasbourg in 1885, then to Zurich and St Petersburg and Paris. In fact, he had already entered early retirement in his fifties when the Russian Revolution erupted, forcing him from his family estate in Ukraine. He joined the Pasteur Institute, where he worked for 24 further years until his death in 1953.

▲ Ukrainian microbiologist Sergei Winogradsky.

Winogradsky lived a rich and prolific life. His discoveries illuminated the nitrogen cycle, the circular flow of nitrogen through the environment that is vital for the healthy growth of all living things. He realized that microorganisms played a huge role in the process, years ahead of his time.

You might imagine, since nitrogen makes up 78 per cent of the Earth's atmosphere, that it would be readily available to plants. But, in fact, atmospheric nitrogen must be turned into a

▲ Nodules on pea roots contain bacteria that fix nitrogen so plants can use it.

usable form for plants to take it up – a process known as nitrogen fixation.

In Winogradsky's time, scientists already knew that pea plants had a cunning way of fixing enough nitrogen to grow. They used bacteria growing

THE NITROGEN CYCLE

Through the nitrogen cycle, nitrogen-fixing bacteria make atmospheric nitrogen available to plants. They, in turn, are eaten by animals.

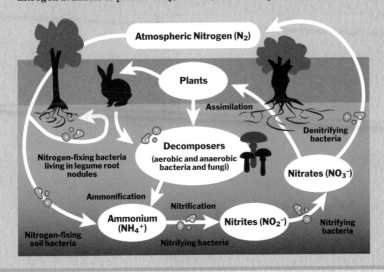

symbiotically in nodules on their roots. In Zurich in 1888, Winogradsky investigated new families of nitrogen-fixing bacteria, and worked out their role in two vital steps for turning nitrogen into the nitrates useful to plants. The process he discovered and named chemosynthesis, in which organisms use chemical reactions instead of sunlight as their energy source, is vital to the nitrogen cycle.

But Winogradsky also isolated

Clostridium pasteurianum, the first microorganism that fixed nitrogen from the air without living symbiotically with a plant. It turned out that 90 per cent of the fixation of nitrogen in the soil is accounted for by such free-living bacteria.

Today, Winogradsky is known as the first microbial ecologist, who saw the matter in which plants grow not as a mass of dead debris but as a living organism. He was a truly scientific son of the soil.

DEADLY CYANIDE

Nitrogen generally exists on an axis between helpful and harmless in everyday life. However, as a component of inorganic cyanide, nitrogen forms part of one of the fastest-acting poisons we know.

A cyanide actually means any chemical compound that contains a cyano group, which consists of a carbon atom triple-bonded to a nitrogen atom. In stable organic cyanides, called nitriles, most of the toxicity is lost and the cyano group forms part of acrylic polymers such as poly(acrylonitrile), superglue and medicinal drugs.

▲ Cyanide crystals look attractive under polarized light, but hydrogen cyanide is lethally toxic.

In hydrogen cyanide, historically called prussic acid, there is a hydrogen attached to the cyano group. This substance is a versatile ingredient in making chemical compounds, including acrylic fibres and plastics such as nylon and perspex. Since the 19th century, salts of hydrogen cyanide such as sodium cyanide and potassium cyanide have also played a huge role in recovering metals in the mining industry.

▲ A thick, colourless liquid, nitrile is involved in synthesizing nylon – here seen in a scanning electron micrograph.

BLOOD AGENT

However, industrial spills and leaks have always made cyanide's use in mining hazardous because of its solubility and toxicity. Cyanide's danger is because it interferes with the cellular process of respiration. A lethal dose will kill within minutes.

Hydrogen cyanide is particularly fearsome because it is a gas at ambient temperatures. It was the chemical used in the horrors of the Nazi gas chambers, and is among the poisons listed as a blood agent among chemical weapons of mass destruction, because it impacts the entire body.

In 1933, American biologist Matilda Moldenhauer Brooks discovered an antidote to cyanide poisoning. Large doses of a medicine called methylene blue could fight the action of the poison if injected. It was a finding that she made through a grant she gained while working at the University of California, Berkeley, and had to defend in print after another scientist published her work as his own. Today, a range of more effective treatments can be used against the effects of cyanide – but only if administered in time.

▲ A scanning electron micrograph of superglue, a cyanoacrylate compound containing a nitrile group.

HOW YOUR CELLS MAKE ENERGY

Nitrogen and phosphorus are key elements in adenosine triphosphate, ATP. This small but powerful molecule transports energy to where it is needed in the cell for functions such as synthesizing proteins or division.

Adenosine is made from the nitrogen-containing base adenine (shown on the right side of the diagram), plus a sugar molecule (in the centre). It is attached to three phosphate groups (on the left).

Fast Facts
PHOSPHORUS P
ATOMIC NUMBER 15

Phosphate rock is the chief source of phosphorus for making fertilizer today. ▶

Character: Phosphorus has several solid forms – allotropes – called red, white, black and violet.

In 1830, French chemist Marc Charles Sauria developed phosphorus-based matches for fire-lighting. They worked well – but were eventually banned because of the toxicity of white phosphorus. Workers inhaling white phosphorus fumes suffered from a serious condition called phossy jaw – a horrible decaying of the jawbone.

During the Second World War, phosphorus bombs were dropped by both sides, and even today they are being used, though with debatable legality.

Discovery: In 1675, German alchemist Hennig Brand imagined that he might find gold in the yellow liquid that is human urine. He reduced 50 buckets of the smelly substance to a paste by evaporation. He then heated it at high temperatures together with charcoal and sand until the fumes began to glow, and sometimes burst into flames. He never found gold – but he did find the first element whose discovery can definitively be credited to a specific person: phosphorus.

Name: *Phosphoros* is the Greek for "bringer of light" and so suited this element well.

World sources: For a hundred years after Hennig Brand discovered it, phosphorus came from processing urine. Then bone was found to be a good source. Today, phosphate rock is the commercial source of phosphorus, with large deposits in Morocco and China.

Incendiaries made of phosphorus were used in the First World War. ▼

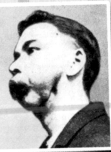

▲ *Workers making matches from toxic white phosphorus suffered a serious condition called phossy jaw.*

▲ Could phosphorus in human urine be the answer to raising crop yields?

URINE AND NATURAL FERTILIZERS

Peak phosphorus might sound an unlikely term for a global crisis. However, it is a possibility that concerns scientists around the world. There is an ever-increasing demand for phosphorus in fertilizer manufacture to increase crop yields, yet a finite supply of phosphorus-rich rocks.

A solution may exist in an unexpected source – urine. Humans excrete all the phosphorus they ingest, and so, before artificial fertilizers and industrial farming, it was the obvious source of soil nourishment.

David Cordell from the Global Phosphorus Research Initiative in Sweden now suggests we make recycling urine a top priority – before farming as we know it goes down the pan.

COCA-COLA AND ITS PHOSPHORIC ACID ADDITIVE

Delicious! Refreshing! Exhilarating! Invigorating! This was how the first Coca-Cola advertisement promoted the fizzy drink to readers of the *Atlanta Journal* in 1886. The drink's inventor was a pharmacist called John Pemberton, who had tried to launch a number of medicines with little success. His formulation for Coca-Cola contained a small amount of coca leaf, the raw material for the manufacture of cocaine. He also included a wonder-drug of the time, the kola nut, which contained stimulants we now know include caffeine.

The drink was intended as a nerve tonic to be sold at soda fountains, good to combat headaches and tiredness. Pemberton added sugar to mask the bitter kola flavour, plus oils including vanilla, nutmeg, orange, lemon, lime and cinnamon. To give the brown colour, he used caramel.

▲ *America's soda fountains began serving Coca-Cola in 1886.*

The drink's potential was clear to Frank Robinson, the man who created its first advertisement and who also came up with Coca-Cola's catchy name and famous logo. When Pemberton died at only 57, Robinson eventually persuaded his new employer, a more successful pharmacist called Asa Griggs Candler, to try the fizzy beverage – and then buy the rights to produce it.

MERCHANDISE NO. 4

Candler and Robinson redeveloped the formula, reducing the cocaine and kola, and incorporating a substance known to other soda-fountain drinks of the period – phosphoric acid, to provide a tart flavour. After adding other ingredients, the pair code-named this flavouring to avoid imitators, and it became *Merchandise No. 4*.

When Robinson came up with novel promotional ideas to get customers asking for Coca-Cola, it began riding a wave of success that has never died down. Coca-Cola's worldwide inundation has not been without controversy, though, and its inclusion of phosphoric acid as an artificial additive met with resistance in several countries.

In 1949, during a campaign against "Coca-Colonization" in France, winemakers and politicians claimed that the addition of phosphoric acid ran contrary to health laws. When a trade war with America loomed, however, the

▲ *Coca-Cola advertisement from the 1890s.*

sparkle went out of the movement, and a judge ruled Coca-Cola legal in France. Today, the ingredient – minus its secretive code-name – is part of the recipe for several Coca-Cola drinks.

Fast Facts

ARSENIC [As]

ATOMIC NUMBER 33

Character: Every kind of arsenic is toxic to humans and animals. Pure arsenic is a shiny grey colour, but it occurs in minerals with vivid colours, such as orpiment, which is yellow-orange, and realgar, which is red. Arsenic is one of the metalloids and has the unusual property of turning straight from solid to gas when heated, a trick known as sublimation.

Discovery: During the Middle Ages, a German philosopher and theologian called Albertus Magnus lived in Cologne. He taught widely, and made contributions to logic, psychology, metaphysics, meteorology, mineralogy and zoology. During his investigations in alchemy, he managed to produce arsenic from orpiment, in around 1250, and is often credited as its discoverer.

Name: Arsenic comes from the Greek word for the yellow pigment, *arsenikon*, made from the mineral orpiment.

World sources: Arsenic is a useful addition to lead to make it stronger, and as an ingredient in semiconductor devices, while in other applications its use is declining because it is so toxic. It is one of the by-products from the purification of other metals in the electrolytic process.

▲ German philosopher Albertus Magnus.

▲ Arsenic occurs in vivid-colour minerals such as orpiment.

► Thrillers and murder stories often feature arsenic poison.

WELL WATER AND ARSENIC

White arsenic is a classic poison in thrillers and murder mysteries. But a tale of real terror is unfolding – and it is due to something in the water. More than 200 million people globally are exposed to dangerous arsenic levels because their drinking water is contaminated by arsenic compounds leaching from the bedrock.

Pollution was such a problem in surface water sources in Bangladesh during the 1970s that the government drilled deep wells. However, one in five of these wells now contains water with arsenic above safe levels. Malnutrition makes arsenic poisoning more serious still.

At its worst, arsenic poisoning causes cancer, brain damage and death. It disrupts the way cells store energy by mimicking phosphates, leaving cells unable to produce energy or send signals – fundamental aspects of staying alive.

Bangladesh may have the most serious levels of arsenic exposure. However, authorities and scientists are tracking arsenic levels in water supplies in the western USA, Chile, Argentina, Romania, Nepal, China, Taiwan, Vietnam and Thailand, where it is also a worry.

▲ In Bangladesh, as well as other countries, arsenic levels in water supplies are a huge health concern.

▼ Rice absorbs high levels of arsenic from the environment in some areas of the world.

Fast Facts
ANTIMONY [Sb]
ATOMIC NUMBER 51

Character: Antimony is a semi-metal and plays a role in electronic devices as a dopant in semiconductors. If processed into a pure form it is hard, brittle and silvery, but is most often found in nature as a mineral called stibnite, antimony sulfide, which forms clumps of elongated crystals.

Discovery: Known in ancient times, antimony was a useful material to the Egyptians and Chaldeans. Antimony helped to harden lead type in early printing presses.

Name: The name antimony derives from Latin *antimonium*, but the source of this word is not clear. It might mean "against aloneness", because antimony is never found by itself in nature. Alternatively, it might come from a mis-transcribed Arabic word *athimar*. The chemical symbol Sb is from the Latin for the stibnite mineral *stibium*.

World sources: The Xikuangshan mine in Hunan, China, is the world's largest source of antimony. It is in the city of Lengshuijiang, which means cold water river. However, many of the mine's mineral deposits were laid down in warm seas millions of years ago.

▲ *Antimony sulfide grows with sword-shaped crystals.*

▲ *Antimony helped harden lead type, giving a cleaner metal casting.*

◄ *The charging efficiency of lead-acid batteries is enhanced using antimony.*

Fast Facts
BISMUTH Bi
ATOMIC NUMBER 83

Character: Bismuth is a silvery brittle metal, tinged with pink, which forms intriguing crystals in its pure form. A tea-table trick popular in Victorian England involved spoons made from an alloy of bismuth, lead and tin with a low melting point. Before guests' astonished eyes, the spoons would disappear into a hot beverage.

Discovery: Often confused with lead, which is similarly dense, bismuth was proven to be a separate metal by Claude-François Geoffroy in 1753.

Name: The word bismuth comes from German *Bisemutum*, a corrupted form of the words "white mass".

World sources: China, Vietnam and Mexico all mine bismuth, but it currently has few commercial applications.

Bismuth alloyed with tin is a non-toxic alternative to lead for shotgun pellets. ▼

BISMUTH CRYSTALS

The stunning staircase shape of bismuth crystals results from the outer edges of the structure growing faster than the insides. An oxide layer on the crystal's surface causes the iridescent rainbow of colours, as different wavelengths of light interfere as they reflect.

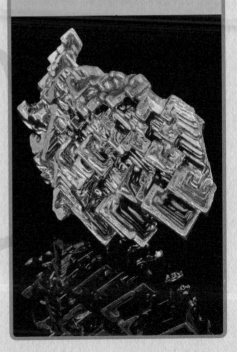

ANTIMONY AND BISMUTH IN THE EGYPTIAN COSMETIC BOX

Men and women alike took care of their appearance in Ancient Egypt. In wall paintings of scenes in daily life, along with the belongings buried with the dead, we have evidence of a thriving interest in make-up.

Both sexes applied rouge to give their cheeks a bloom, plus alluring red ointment as lipstick. They made green eyeshadow from malachite – copper carbonate – and later favoured grey shades made from galena – lead sulfide.

Antimony sulfide was a constituent of kohl, which men and women used as a black or grey eyeliner to imitate the markings of the sun god Horus. They also applied antimony sulfide as a mascara. Kohl was not just worn for its dramatic eye-opening look. The Egyptians thought it repelled flies and protected your eyes from both bright sunlight and infection.

Bismuth also helped the Egyptians to shine – in the form of bismuth oxychloride, it made cosmetic products glitter. Today, it is still a component of pearlized products from eyeshadow to nail varnish.

▲ Ground-up galena, or lead sulfide, made grey eyeshadow for Ancient Egyptians.

◄ Eyeliner made with antimony sulfide was a popular choice. As well as looking dramatic, people believed it protected the eyes.

MOSCOVIUM Mc
ATOMIC NUMBER 115

Character: Highly radioactive, moscovium may be a dense, metallic solid – but no one is sure. It is one of the super-heavy elements not found in nature.

Discovery: An international team created element 115 in 2003 at the Joint Institute for Nuclear Research (JINR) in Dubna, Russia. Led by Yuri Oganessian of JINR and Ken Moody of the Lawrence Livermore National Laboratory, California, the collaboration had yielded elements 114 and 116 in 1998.

Name: Element 115 was named after the city of its discovery, Moscow.

World sources: Only a few dozen atoms of moscovium have ever existed.

▲ *Egyptian men and women wore make-up to enhance their appearance.*

CHAPTER 10

THE OXYGEN GROUP

WHY MIGHT NAKED MOLE RATS HAVE CLUES
FOR SURVIVING A HEART ATTACK?

WHAT ARE SULFUR AND OXYGEN'S ROLES IN THE VIBRANT
COLOURS OF CHINA'S RAINBOW MOUNTAINS?

HOW DID THE HAYA PEOPLE OF TANZANIA
MAKE STEEL 1,500 YEARS AGO?

THE OXYGEN GROUP

Elements in the oxygen group gather under the name chalcogens, a term they were given in 1932 by German chemists Wilhelm Blitz and Werner Fischer. It means ore-former, from the Greek for ore, *chalcos*, because they tended to appear in copper ores.

All the oxygen group elements have six outer electrons, making them highly reactive. Except for oxygen, they are all solid at room temperature, and are all non-metals or metalloids. Livermorium may be a metal, but only a few atoms have ever existed.

FAREWELL TO PHLOGISTON

Elizabeth Fulhame produced only one scientific publication in her lifetime, but it was a bold one. Her 1794 essay made the statement that both the phlogiston and anti-phlogiston theories were wrong.

Phlogiston was a theoretical substance, first suggested in 1667, that tried to explain combustion. When something burned in air, the idea went, it released phlogiston which the air would absorb. Leading scientists including Joseph Priestley pursued this theory. In his experiments on gases in the early 1770s, Priestley called one gas "dephlogisticated air", because he thought its ability to encourage burning was a sign it could readily absorb phlogiston.

But Antoine Lavoisier had increasing doubts about phlogiston, because he found that metals often gained mass during combustion reactions – contrary to the idea that they had lost phlogiston.

◀ *In the 18th century, many scientists held that a mysterious substance called phlogiston was released during burning.*

▲ *English chemist Joseph Priestley isolated oxygen but called it "dephlogisticated air" because it encouraged burning.*

He proposed that the metals were combining with a component of the air, which he realized was Priestley's gas. In 1778, he named it oxygen.

In her publication, Elizabeth Fulhame rejected the phlogiston theory, and challenged aspects of Lavoisier's anti-phlogiston findings. Her confidence was based on 14 years of work that Priestley himself had encouraged her to write up – primarily experiments in which she produced metals from their salts. To her, water was key for all reduction and oxidation.

Fast Facts
OXYGEN $\boxed{\text{O}}$
ATOMIC NUMBER 8

Character: The most abundant element in the Earth's rocky crust, oxygen also makes up a fifth of the atmosphere as a transparent, but vital, gas. Another form of oxygen, the allotrope ozone (O_3) in the upper atmosphere, protects all living organisms from harmful UV radiation emitted by the Sun.

Discovery: Many chemists produced oxygen without realizing their success, because all assumed that air was a single element rather than a mixture. However, in 1774 in England, Joseph Priestley revealed he had decomposed mercuric oxide and produced a gas that caused a candle to burn more brightly. Carl Wilhelm Scheele had done a similar experiment in Sweden in 1771 but had not yet published it. When French chemist Antoine Lavoisier also produced the gas, he realized it was key to understanding combustion.

Name: Lavoisier named oxygen from the Greek word *oxys* meaning sharp, because he believed, wrongly, that oxygen was a part of all acids.

World sources: Industries extract 100 million tonnes of oxygen annually through the process of fractional distillation of liquid air. Fifty-five per cent goes into steelmaking, and 25 per cent to industrial chemical production.

History showed that Lavoisier was right. Phlogiston theory was snuffed out and oxygen theory took its rightful place. Fulhame overstated the importance of water – but three claims to fame do emerge from her single publication.

First, she showed that metals could be produced by aqueous chemical reduction at room temperature, rather than by high-temperature smelting. Second, she realized that water enabled many reactions but was recovered unchanged afterward – the earliest-known mention of a catalyst. Third, she used light to reduce silver salts, the chemical basis for the photographic process.

You could add a further achievement, which is that through her one work, which became famous in America and was translated into German, Fulhame can today be identified as the first solo female researcher in modern chemistry.

▲ Antoine Lavoisier worked alongside his wife, Marie-Anne Paulze Lavoisier, who played a crucial supporting role in many chemistry experiments.

AN

ESSAY

ON

COMBUSTION,

WITH A VIEW TO A

NEW ART

OF

DYING AND PAINTING.

WHEREIN

THE PHLOGISTIC AND ANTIPHLOGISTIC HYPOTHESES
ARE PROVED ERRONEOUS

By Mrs. FULHAME.

LONDON:

PRINTED FOR THE AUTHOR,

BY J. COOPER, BOW STREET, COVENT GARDEN,

And Sold by J. JOHNSON, No. 72, St. Paul's Church Yard ;
G. G. and J. ROBINSON, Paternoster Row ; and
T. CADELL, Jun. and W. DAVIES, Strand.

1794.

[ENTERED AT STATIONERS HALL.]

Elizabeth Fulhame published a scientific work in 1794 in which she used her experimental results to challenge views on combustion. ▶

OXIDATION AND REDUCTION REACTIONS

Oxygen gas held the key to understanding combustion, and gave its name to a variety of chemical processes called oxidation. Through Antoine Lavoisier's work, reactions in which oxygen was gained became classified as oxidation, while those in which oxygen was lost were called reduction. Elizabeth Fulhame recognized that these were reversible reactions and complementary processes, today known as redox reactions:

OXIDATION

When a hydrocarbon fuel burns, oxygen is gained:

hydrocarbon (e.g. methane) + oxygen (from the air) → carbon dioxide + water

Or when a metal such as magnesium burns in air it is oxidized or loses electrons:

magnesium + oxygen → magnesium oxide

And rusting is also an oxidation reaction that requires both air and water:

iron + water + oxygen → hydrated iron(III) oxide (rust)

REDUCTION

Metal oxides can be reduced (losing oxygen or gaining electrons) to form metals using a reducing agent such as hydrogen:

copper oxide + hydrogen → copper + water

The process of photosynthesis reduces carbon dioxide using the energy in sunlight, reversing the oxidation performed by the atmosphere on substances in the Earth's crust:

carbon dioxide + water → glucose + oxygen

A BREATH OF AIR

With every breath, you take in oxygen to supply the process of respiration. The oxygen reacts with glucose in your cells to make carbon dioxide and water – and crucially, releases energy. Without oxygen, you, like any other living creature, would die.

At least, that is what biology knew until experiments with naked mole rats. These strange-looking, long-living rodents have all kinds of interesting abilities, including being immune to the irritant in chillies, and never getting tumours.

In 2017, researchers in America and Germany found that the rats could function perfectly well in an atmosphere containing only 5 per cent oxygen instead of the usual 21 per cent. When the oxygen levels were reduced to

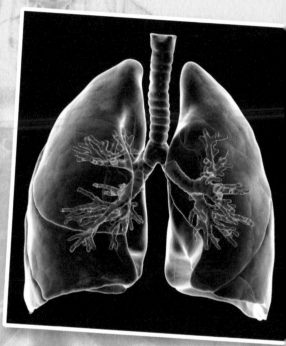

▲ Your heart pumps blood around your body, and your blood is oxygenated in the lungs.

zero, the creatures went into a kind of suspended animation, and their heart rates dropped from 200 beats per minute to around 50. Even with a wait of 18 minutes before normality was restored, the rats recovered without harm.

METABOLIC SHIFTS

So far, so good. Just as in humans, haemoglobin carries oxygen around naked mole rats' bodies. Other findings had already showed the rats' version of the molecule was particularly efficient, possibly helping them survive in oxygen-poor nests deep underground.

But something more incredible was happening, as analysis of tissue from the rats' brains and hearts revealed. In the absence of oxygen, the animals had used fructose instead of glucose to produce energy to run their vital processes – just as plants do.

The scientists wonder if a similar metabolic shift could be triggered in humans as a back-up system to avoid the kind of oxygen starvation damage seen after strokes and heart attacks.

▶ Naked mole rats are accustomed to surviving in low-oxygen conditions in underground nests. In the absence of oxygen, naked mole rats use fructose for energy – like plants.

STEELMAKING AND OXYGEN

In 1855, an English inventor called Henry Bessemer discovered the process of blowing air through molten pig iron, making it burn white-hot. He succeeded in removing impurities from the metal and produced very strong steel – an alloy of iron and carbon.

The Industrial Revolution was steaming ahead in England, but its reliance on cast iron was an increasing problem. Used in structures since the 1770s, cast-iron beams had failed and caused a number of fatal bridge collapses through its brittleness and high rate of rusting. Bessemer's steel promised to be a cheap replacement for wrought iron that was the only alternative – both in civil engineering and his particular interest, weapons manufacture.

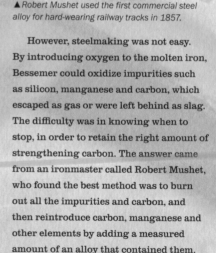

▲ *Robert Mushet used the first commercial steel alloy for hard-wearing railway tracks in 1857.*

◄ *Henry Bessemer discovered the steelmaking process in 1855.*

However, steelmaking was not easy. By introducing oxygen to the molten iron, Bessemer could oxidize impurities such as silicon, manganese and carbon, which escaped as gas or were left behind as slag. The difficulty was in knowing when to stop, in order to retain the right amount of strengthening carbon. The answer came from an ironmaster called Robert Mushet, who found the best method was to burn out all the impurities and carbon, and then reintroduce carbon, manganese and other elements by adding a measured amount of an alloy that contained them.

IRON AND STEEL

Pig iron that has just been smelted in a furnace contains 4 per cent carbon along with nitrogen and silicon impurities. It is very brittle.

By lowering the carbon content of iron, and controlling other impurities, iron turns to **steel**. In the blast furnace process used today, oxygen is pumped into the molten mixture, and impurities like carbon, phosphorus and nitrogen are oxidized and removed.

By this process, Mushet made the first railway tracks from steel, laid at a busy station in 1857, which proved incredibly hard-wearing. He applied for a patent for his tungsten-strengthened steel, the first commercial steel alloy, but was not such a good businessman or entrepreneur as Bessemer, whose patents and processes ensured relentless success.

ANCIENT FURNACES

While steel production enabled industry to spread all over the world in the 19th century, its story actually has an intriguing early chapter. In the late 1960s, anthropologist Peter Schmidt was studying the complex heritage of the Haya people in northwestern Tanzania. He realized that in their oral tradition, passed down the generations, were stories of steelmaking.

In 1976, at Schmidt's request, the oldest among the Haya recreated a traditional furnace, a cone-shaped structure built over a pit packed with charred reeds. The reeds provided carbon, which combined with iron ore to produce steel. Key to the process was pumping preheated air by hand into the charcoal-powered furnace, producing hotter temperatures than anything known in Europe before Bessemer.

The following year, Peter Schmidt located excavated remains of more than a dozen ancient furnaces as described by the Haya, on the shores of Lake Victoria. Radio-carbon dating showed they were between 1,500 and 2,000 years old.

Fast Facts

SULFUR $\boxed{\text{S}}$

ATOMIC NUMBER 16

Character: One of only a few non-metal elements that exist in pure form in nature, sulfur is a yellow crystalline substance. It is also a common component of minerals including celestine, which is sometimes a limpid blue.

Discovery: All living things contain sulfur and it has been a useful natural substance for centuries. Chemist Humphry Davy thought it contained hydrogen, but his sample turned out to be contaminated. In 1809, its status as an element was confirmed by French scientists Joseph Gay-Lussac and Louis-Jacques Thénard.

▲ *Iron disulfide is also known as iron pyrite or fool's gold.*

Name: The word sulfur comes from its Latin equivalent *sulpur*.

World sources: Sulfur has long been associated with hellish scenes and ideas of fire and brimstone. But it also helped create an incredible striped landscape of yellow iron sulfide and red iron(III) oxide in the Zhangye Danxia mountains in China.

▼ *The Rainbow Mountains with their incredible colouring are in the Zhangye Danxia Landform Geological Park in Gansu, China.*

THE BEAUTY AND HORROR OF NATURAL SULFUR

Volcanoes are the natural habitat of the element sulfur. At Indonesia's Kawah Ijen volcano in Java, sulfur-rich gases escape from the rock, igniting with an eerie blue flame. They stream down the volcano's slopes at night, sending flames 5m (16.4ft) into the sky. Visitors love to see this nightly light show – although from a distance, as the gas reacts with water vapour and rains sulfuric acid onto the surroundings.

Tourists also come to Kawah Ijen to see its milky turquoise lake, stunningly beautiful but dangerously acidic, and the largest of its kind in the world. Yellow sulfur is deposited around the shore, and the area is shrouded in a sulfurous cloud.

But trippers are not the only volcanic visitors. Hundreds of local miners gather daily to make the descent into the crater. With only basic equipment and lacking protective clothing, they break sulfur slabs formed there, and carry loaded baskets up to the surface for processing. Several deaths have resulted from toxic fumes, and life expectancy is low. Nonetheless, many miners hope the money they earn will mean education for their children so they may rise above this way of scratching a living from the Earth.

▲ Kawah Ijen has an acidic lake with sulfur deposits at its shoreline. Miners carry yellow sulfur slabs from within the volcanic crater.

ELEMENTS THAT SMELL

Our sense of smell is powerful and primeval – there are aromas that make our mouths water and odours that make us feel sick. Yet the way smell works is a young science. American scientists Richard Axel and Linda Buck shared a Nobel Prize in 2004, which recognized their major discovery of odorant receptors, and the organization of our olfactory system.

▲ *Garlic, grapefruit and coffee aromas contain sulfurous compounds called thiols.*

Among the elements that get our smell receptors complaining is sulfur, particularly when bonded to hydrogen in organic compounds, making substances called thiols. Our noses can detect thiols at tiny concentrations. They provide aspects of some very pleasant smells, like roasted coffee and grapefruit. But they also make their presence known in garlic, rotting food, skunk spray and human sweat. An additive made of thiols in natural methane makes the otherwise odourless gas easier to detect in case of leaks.

Sulfur is not the only pungent member of this group. Selenium is a vital trace element for our health, but if it replaces sulfur in a volatile substance, the smell can be even more intense. People who work with the chemical take enormous care not to come into skin contact with it, as it causes horrible body odour and bad breath. Tellurium also has the delightful property of causing awful breath because the body converts tellurium to dimethyl telluride. This chemical smells like garlic – but garlic that has rotted.

HOW DOES YOUR NOSE SMELL?

Smell relies on molecules of a substance reaching the back of your nose, where they meet the olfactory epithelium. This strip of mucus-covered tissue has millions of sensory nerve cells, which contain about 450 types of olfactory receptors. When an odour molecule activates a receptor, an electrical signal travels from the nerve cell to the olfactory bulb, a structure in the brain that sends the information on to other brain areas for processing.

But how can we smell so many different scents? Each receptor can be activated by many different smell molecules – and each smell molecule can activate several receptors, all to varying degrees, creating a kind of smell code. What we experience as a single smell is a complex combination of molecules acting on a variety of receptors.

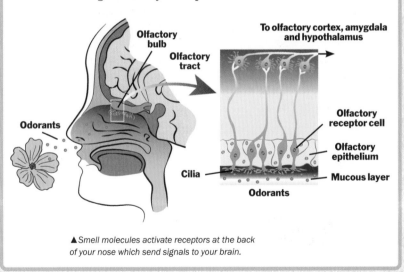

▲ Smell molecules activate receptors at the back of your nose which send signals to your brain.

Fast Facts
SELENIUM $\boxed{\text{Se}}$
ATOMIC NUMBER 34

Character: Selenium has two main solid allotropes, a pure grey form, and a red form that is soft and powdery. Its main industrial application is in colouring glass and ceramics, and it has a vital role as a micronutrient in the body.

Discovery: Swedish chemist Jacob Berzelius discovered selenium in a red sediment that collected at the bottom of vessels holding sulfuric acid. Initially he thought it was tellurium because it had similar properties, but realized it was a distinct element.

Name: Berzelius named selenium after the Greek word for the Moon – *selene*.

World sources: Most of the world's selenium production comes as a by-product of refining sulfide ores of metals including copper, nickel and lead. Germany and Japan are the top producers.

▲ *Selenium, discovered in 1817, is a non-metal, but one of its allotropes is grey, crystalline and metallic-looking.*

Fast Facts
TELLURIUM $\boxed{\text{Te}}$
ATOMIC NUMBER 52

Character: In its pure form, tellurium is a brittle, silver-white metalloid. It is mildly poisonous, but its more interesting physical effect is to give people body odour and bad breath, even at extremely low concentrations.

Discovery: The gold-mining town of Zlatna in Romania was the original source of tellurium. Austrian mining engineer Franz-Joseph Müller collected a mineral there which he initially thought was antimony because of its metallic look, but turned out to be a combination of gold and a mystery element. He puzzled over it for three years, calling the mineral his *metallum problematum* and eventually decided it was indeed a new element.

Name: *Tellus* is the Latin word for Earth, and German chemist Martin Klaproth gave tellurium this name after Müller sent him a sample.

World sources: Tellurium is a by-product of lead and copper refining carried out in the USA, Japan and Canada.

▲ The original sources of tellurium were gold mines in the Romanian town of Zlatna.

▲ Tellurium is a semi-metal, brittle and silver-white.

▲ Austrian engineer Franz-Joseph Müller identified tellurium in 1783 in a mystery mineral he found in Romania.

STELLAR ELEMENTS

▲ *Chemically similar, tellurium and selenium took their names from words for Earth and Moon.*

Selenium and tellurium have names inspired by the Moon and Earth. But to see where they came into being, you have to look to the stars.

The Big Bang produced hydrogen and helium in abundance, and a little lithium too. All the heavier elements are a result of star factories labouring over the 13.7 billion years since. Our Sun can manufacture elements, but only up to the mass of oxygen. After that, bigger, more energetic, stars are required to fuse protons together to make heavier nuclei. In 2012, astronomers announced that they had seen selenium and tellurium in a 12-billion-year-old star within the halo surrounding the Milky Way.

So how does a star actually form elements like selenium and tellurium?

Most elements heavier than iron come into being through neutron-capture nucleosynthesis, which can take place during the violent supernova death of a star. Neutrons decay into protons within a nucleus, producing elements with larger atomic numbers. It all happens so fast that it is known as the rapid process.

The newly minted elements are then spread into space as gas clouds by the supernova explosion – where they eventually collapse in again, forming new stars containing the heavier elements, and start the process again.

Fast Facts
POLONIUM Po
ATOMIC NUMBER 84

Character: Polonium is a silvery-grey metalloid that self-heats because of its intense radioactivity. It is extremely rare in nature.

Discovery: Marie and Pierre Curie discovered polonium in 1898, extracting it from tonnes of uranium ore, pitchblende, that they processed by hand.

Name: Marie Curie named polonium after her country of birth, Poland, which was at the time occupied by Germany. It is one of only a handful of elements named after countries.

World sources: Russia makes all the commercial polonium in the world. Its pioneering lunar rovers, Lunokhod 1 and 2, used polonium as a heat source during their missions in the 1970s.

Fast Facts
LIVERMORIUM Lv
ATOMIC NUMBER 116

Character: Highly radioactive, livermorium was first made in the year 2000 at the Joint Institute for Nuclear Research (JINR), in Dubna, Russia.

Discovery: In the successful experiment, scientists in Russia used calcium atoms to bombard a curium target supplied by the Lawrence Livermore National Laboratory in California. They produced a single atom of a new element, which decayed by emitting alpha radiation.

Name: The name livermorium for element 116 was officially recognized at the same time as that of flerovium, element 114, marking a Russian-American collaboration that had yielded both new substances.

World sources: Four isotopes of livermorium are now known to exist, but the longest-lasting has a half-life of around 60 milliseconds.

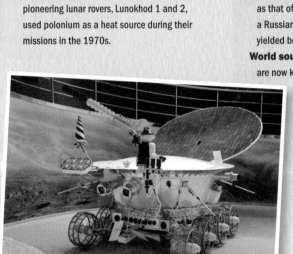

◄ A model and prototype of Lunokhod 1, the first remote-controlled lunar roving vehicle, launched in 1970 by the Soviet Union.

9
F
Fluorine

17
Cl
Chlorine

35
Br
Bromine

53
I
Iodine

85
At
Astatine

117
Ts
Tennessine

CHAPTER 11

THE HALOGENS

WHY DID SO MANY SUFFER
IN THE SEARCH FOR FLUORINE?

HOW DID THE COVENTRY REMEDY HELP SAVE LIVES?

WHAT IS THE SCIENCE BEHIND
THE ART OF PHOTOGRAPHY?

THE HALOGENS

The group with fluorine at the top is often known as the halogens, meaning *salt-former* in Greek. The halogens are all non-metals, and produce salts when they react with metals, giving us sodium chloride, calcium fluoride, potassium bromide, silver iodide and many more.

All the elements in this group are missing one electron to complete their outermost valence shell, making them highly reactive. At the top of the group, fluorine is an extremely dangerous element that burns skin and lungs. You do not want to breathe in chlorine gas either, and bromine and iodine are also toxic, while astatine is highly radioactive. However, your body contains quantities of the first four halogens, and iodine is essential in your daily diet to supply the thyroid gland.

The halogens are the only chemical group to contain elements that exist in the three primary states of matter if you encounter them at room temperature. Fluorine and chlorine are gases, bromine is a liquid, and iodine and astatine are solids. Only a tiny quantity of tennessine has ever been made, so you are unlikely to experience it at any temperature.

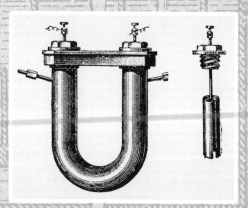

◀ *Henri Moissan's apparatus for the preparation of fluorine.*

Fast Facts
FLUORINE [F]
ATOMIC NUMBER 9

Character: Fluorine is the most reactive of all the elements in the periodic table. It is a pale yellow gas and only helium, neon and argon can resist its advances in terms of bonding.

Discovery: Fluorides were useful minerals before anyone knew what fluorine was. Fluorspar, which is calcium fluoride, made a good flux for keeping things flowing during smelting by reducing the melting point. But fluorine itself was elusive and incredibly hard to isolate. The first chemist to manage it was the French chemist Henri Moissan, who won the Nobel Prize for Chemistry in 1906 for his achievement.

Name: Chemists in the early 19th century anticipated that an element similar to chlorine must exist, and André-Marie Ampère suggested the name fluorine, recalling fluorspar and the Latin *fluere*, to flow. Although he later preferred the name phthorine, from the Greek *phthoros* meaning destructive, fluorine stuck.

World sources: Fluorspar is still used in refining metals and there are large deposits in South Africa, Mexico and China. Hydrogen fluoride, from which fluorine gas is produced, is a by-product of manufacturing phosphoric acid.

▼ *Sodium fluoride in toothpaste or drinking water can prevent dental decay.*

▲ *Fluorspar, or fluorite, is an important industrial mineral, but crystals are sometimes also polished as gemstones.*

▶ We still produce fluorine gas using the method Henri Moissan devised, although it can now be stored and transported much more safely.

DANGER ELEMENT

Fluorine took so many swipes at scientists trying to find it that there is an unfortunate group known as "fluorine martyrs". Belgian chemist Paulin Louyet was one. In 1850 he died from inhaling hydrogen fluoride vapours. Jerome Nicklès, who taught chemistry in Nancy, France, succumbed to the same unforgiving fumes in 1869.

Fluorine had no respect for fame, either. Renowned scientists Humphry Davy, Joseph Louis Gay-Lussac and Louis Jacques Thénard all suffered injuries as a result of experiments with fluorine during their careers.

Davy's experiment with fluorine left him with eye damage. But his work in separating other elements using electrolysis had been highly successful. Two brothers, George and Thomas Knox of Ireland, tried Davy's approach with fluorine in 1836. They passed electricity through a molten metallic fluoride salt but ended up with dreadful results –

one nearly died, and the other was unwell for three years.

It was almost as if fluorine did not want to be found. English chemist George Gore successfully isolated a small amount of fluorine gas in 1860 – whereupon an explosion destroyed part of his laboratory.

No wonder that when Henri Moissan won the Nobel Prize in 1906 for discovering fluorine, the committee praised his skill in studying what they referred to as a savage beast among the elements.

Fast Facts
CHLORINE `Cl`
ATOMIC NUMBER 17

Character: A greenish-yellow gas, chlorine is very
reactive and does not occur as a pure element in
nature. However, it is abundant in the Earth's crust
as salt, and in seawater.

Discovery: Swedish chemist Carl Wilhelm Scheele
studied chlorine gas in 1774 and is credited with
chlorine's discovery, although he thought it was a
compound. He called it "dephlogisticated muriatic
acid air" because he made it from hydrochloric
acid – then known as muriatic acid. In 1810,
Humphry Davy had another look at the experiment,
and correctly decided chlorine was an element.

Name: *Chloros* means pale green in Greek and led
to the name Humphry Davy gave the element in
recognition of its colour.

World sources: The world's great salt producers
are China, USA, India, Canada, Brazil and
Australia, but almost all countries have some
manufacturing capacity.

▼ *Chlorine doesn't occur naturally in a pure
form but is common as sodium chloride – salt.*

▼ *Chlorine in swimming-pool water kills
bacteria that may be dangerous to health.*

▲ *Manufacturing the plastic polyvinyl chloride
(PVC), with hundreds of everyday uses, accounts
for around a quarter of all chlorine produced.*

CFCs AND THE OZONE HOLE

Non-toxic, non-corrosive and non-flammable – chlorofluorocarbons sound like such a nice group of chemicals. They first made an appearance as refrigerants in the 1930s and also proved useful as propellants in aerosols. The great thing was, CFCs were so inert, it seemed as if they would simply sit in the atmosphere, not reacting, not dissolving, when released. There was not a cloud in the sky.

By the 1970s, the forecast was not looking as bright. Scientists began to realize that CFCs did indeed break down in the intense solar radiation in the stratosphere. This released highly reactive chlorine atoms which attacked the ozone layer protecting us from UV radiation. CFCs were also powerful greenhouse gases, trapping radiation and warming the lower atmosphere.

Within a couple of generations, CFCs had gone from everyone's favourite chemical to social pariah. In 1987, under the Montreal Protocol, international agreements were made to phase them out.

▲ *Mario Molina realized refrigerators and deodorants contained dangerous chemicals.*

SPEAKING POLLUTANT TRUTH TO POWER

Imagine making a computer model of a well-known commercial chemical and its action on the atmosphere. Then think about what it would be like to realize that your model shows that the chemical – used in your fridge, the insulation in your home, even in your deodorant – is destroying the protective layer of ozone around the Earth, just a few millimetres thick.

If you were a young postdoctoral student, what would you do? Mexican-born chemist Mario Molina faced this issue while working at the University of California, Irvine, in 1973. He swallowed his nerves, and took his research to his

CHLOROFLUOROCARBONS

According to Mario Molina's 1973 model, CFC gases rise into the stratosphere, where UV radiation breaks them into their component elements of chlorine, fluorine and carbon.

$$CFCl_3 \xrightarrow{\text{UV radiation}} CFCl_2 + Cl$$

$$CFC_2Cl_2 \xrightarrow{\text{UV radiation}} CF_2Cl + Cl$$

Chlorine reacts with ozone (O_3) to form oxygen.

$$Cl + O_3 \rightarrow ClO + O_2$$
$$ClO + O_3 \rightarrow Cl + 2O_2$$

overall $\qquad 2O_3 \rightarrow 3O_2$

In 1986, American chemist Susan Solomon provided the theory for why ozone was depleted at the poles, when CFCs were used so far away. Polar stratospheric clouds, commonly over Antarctica, attracted the CFC molecules electrostatically, where the ozone-depleting reactions could take place on the solid surfaces of ice crystals.

▲ *Frank Sherwood Rowland and Mario Molina showed chlorofluorocarbons were damaging the Earth's ozone layer.*

boss, Frank Sherwood Rowland, who recognized its importance. They published a paper in 1974, which showed that if CFC release continued, the ozone layer would soon be significantly depleted.

The response was immediate from people concerned about the environment, and CFCs were soon restricted. Some, however, criticized Molina and Rowland's calculations.

Then in 1985 came a second moment for against-the-odds science. Joe Farman was a physicist working for the British Antarctic Survey. He had been collecting atmospheric data in Halley Bay, Antarctica, for years, relying on traditional

devices such as weather balloons and an old ozone-measuring meter. In 1982 and 1983, his ozone readings suddenly showed a shocking drop.

The trouble was that NASA, with a fleet of hi-tech satellites monitoring the atmosphere, had not picked up any similar changes in ozone levels. One of Farman's superiors cautioned that they should not share the data in case it proved embarrassingly wrong.

But Farman and his colleagues courageously published their findings in 1985, revealing that ozone above the Antarctic had fallen by about 40 per cent between 1975 and 1984. They could

▲ *NASA monitors Earth's climate using a network of satellites.*

Character: Bromine is the only non-metal to be liquid at room temperature, forming an orange-brown fluid. It is reactive enough not to appear in pure form in nature, and its compounds easily dissolve in water.

Discovery: Carl Löwig was a student at Heidelberg University in Germany when he managed to produce bromine from the waters of a spring near his home in 1825. However, another young scientist, French chemist Antoine Balard, distilled bromine from a solution of seaweed ash and published his findings first.

Name: Bromine is named after the Greek word for stench, because of its characteristically bad-smelling vapours.

World sources: Brine and salt lakes are the main sources of bromine, which is manufactured in the USA and Israel.

show that CFCs were to blame. NASA rechecked the satellite data and confirmed Farman's findings.

Farman, along with many other scientists, made their voices heard in the environmental debate that has shaped modern times. In 1995, Molina and Sherwood, along with Swedish scientist Paul Crutzen who had worked on the chemistry of nitrogen oxides and ozone, shared the Nobel Prize for Chemistry for their work on ozone and atmospheric chemistry.

▶ *Bromine forms an orange-brown liquid at room temperature.*

▲ *Brine lakes supply bromine commercially.*

WRITING WITH LIGHT

Over a thousand years ago, an Arab scholar called Ibn al-Haytham wrote a treatise in which he described experiments with a camera obscura. In a dark chamber, light shines in through a tiny hole, and forms an upside-down image on the opposite wall. He realized that the device showed that rays of light travel in straight lines, and included this in his seven-volume *Book of Optics*. Among his other findings was the revelation that we can see things because light reflects off objects and into our eyes.

▲ *Ibn al-Haytham wrote a huge* Book of Optics *in the 11th century.*

In his practical experiments and theories, Ibn al-Haytham was continuing a tradition, begun by thinkers including Euclid and Ptolemy, to apply geometry to the problem of explaining vision. Through the Latin translation of his book, his ideas influenced scientists who came after him, both during Islamic civilization and during the European Renaissance.

But if earlier thinkers brought together theories of light with mathematics, the 19th century saw a marriage of light and chemistry. In about 1717, a German physics professor called Johann Heinrich Schulze had observed a strange phenomenon. A bottle of silver nitrate and chalk that he left on a windowsill turned dark in the sun – except from a line left where a curtain

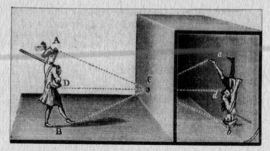

▶ *A camera obscura shows that light travels in a straight line.*

cord blocked the light. He experimented to confirm that it was light rather than heat that had the effect, and even cut out stencils to create light-darkened words on his jar of silver salts.

Schulze's investigations went no further. But in the 1800s, Thomas Wedgwood, son of the founder of the Wedgwood ceramics company, began trying to capture an image in a camera obscura, a project in which he collaborated with the chemist Humphry Davy. Their camera experiments were unsuccessful, but they did manage to produce silhouettes and shadows on a sheet sensitized with silver nitrate. With no way of stopping the process, however, the surface soon turned dark all over.

FIRST PHOTOGRAPHS

Enter Joseph Nicéphore Niépce, a French scientist and inventor. He discovered a contact printing process using a light-sensitive resin called bitumen of Judea. It hardened on a metal plate when exposed to light, allowing the remainder to be washed away and the plate etched with acid and used for printing. It worked very well for reproducing line drawings, but something more sensitive was needed for real-life scenes. From 1828 he applied the bitumen resin to a silver plate, exposed it to bright sunlight and dissolved the

unexposed bitumen as before. Then he covered the plate with iodine fumes. The iodine oxidized the silver where it was not protected by the resin, creating a layer of silver iodide that would blacken in the light. Niépce had created the first black-and-white photographs on metal plate.

▼ *The oldest-known photograph, taken by Joseph Nicéphore Niépce.*

▲ *Joseph Nicéphore Niépce.*

The following year, Niépce entered into an agreement with theatrical artist Louis-Jacques-Mandé Daguerre to perfect the practical aspects of the invention and share any profits. Sadly, Niépce died in 1833, leaving Daguerre to continue the research alone and announce the daguerreotype in 1839. This process for making images on silver plates instantly became hugely popular.

In England, William Henry Fox Talbot had been trying to capture images too – largely because his efforts at hand-drawing had proved frustratingly poor.

▲ Louis-Jacques-Mandé Daguerre.

LIGHT-SENSITIVE SALTS

Silver chloride (AgCl) quickly darkens on exposure to light by disintegrating into elemental chlorine and metallic silver.

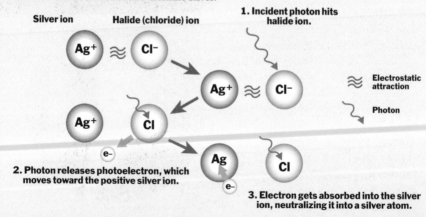

Silver ion

Halide (chloride) ion

1. Incident photon hits halide ion.

≈ Electrostatic attraction

Photon

2. Photon releases photoelectron, which moves toward the positive silver ion.

3. Electron gets absorbed into the silver ion, neutralizing it into a silver atom.

◀ William Henry Fox Talbot wanted to perfect photography because he was bad at drawing.

▲ Made from the earliest camera negative, this photograph was taken by William Henry Fox Talbot at his home, Lacock Abbey.

He devised photogenic drawing in 1834, in which he used paper coated with salt solution and soaked in silver chloride. He placed a piece of lace or a botanical specimen on the paper and exposed it to sunlight, leaving a silhouette image that he fixed with potassium iodide. In 1835, he turned his attention to camera obscura images, succeeding in creating negative images of parts of his house.

Astronomer John Herschel – who came up with the term photograph, meaning writing with light – invented a new fixing method, which Talbot immediately took up, as did Daguerre. This was sodium thiosulfate, or hypo, which was effective in washing away the unexposed silver salts.

Talbot then discovered a method for producing a latent image with a short light exposure, which could be developed into a negative and fixed with hypo. The benefit was that the negative could produce multiple positive prints instead of the single plate of a daguerreotype. Talbot patented the calotype process, named after the Greek words for beautiful impression, in 1841.

◄ Fern fronds captured using photogenic drawing by William Henry Fox Talbot.

developer, Talbot would see a range of depths and densities of shade appearing where light had hit the paper.

The calotype could then be rinsed and stabilized – either with hypo, or with a solution of potassium bromide that converted the remaining silver iodide into silver bromide.

To make copies from the negative calotype image, Talbot used the light-sensitive paper he had previously invented for photogenic drawing. The negative image was sandwiched under glass, on top of a sheet of the photogenic paper, and exposed to bright light. After about 15 minutes, a visible image formed that could be fixed. Today, these are known as salted paper prints.

CHEMISTRY OF CALOTYPES

To create a calotype, Talbot first created iodized paper by brushing it with silver nitrate, drying it, then dipping it by candlelight into potassium iodide, washing, and drying it again. This created silver iodide on the paper.

Before placing the paper into the camera, Talbot brushed the sheet again, with a solution containing silver nitrate and acetic acid. Once exposed to light in the camera, a latent image formed on the paper. Talbot then developed the image by applying more of the solution, plus gallic acid, while warming the paper. As the silver compounds responded to the

▲ Fox Talbot invented the first negative-positive process in photography, creating prints called calotypes.

Fast Facts

IODINE [I]
ATOMIC NUMBER 53

Character: Iodine is the only member of the halogen group that is solid at room temperature. It is a shiny, violet-black non-metal that turns straight into a purple gas when heated in the process called sublimation.

Discovery: Smelly heaps of rotting manure, animal offal and soil played a redolent role in the discovery of iodine. They were used in the manufacture of saltpetre which, from 1803 onward when France was at war, helped supply the French army with gunpowder. But the other vital ingredient was potassium, which Paris-based chemist Bernard Courtois made from seaweed ash. One day, while investigating corrosion in his reacting vessels, Courtois added sulfuric acid and saw a purple vapour, which he found he could condense into dark crystals with a metallic gleam.

Name: Fellow French chemist Joseph Louis Gay-Lussac confirmed that the new find was an element and named it *iode*, iodine in English, from the Greek word for violet.

World sources: Iodine is the least common of the stable halogens. Minerals extracted in Chile are among the richest in iodine, while Japanese and US mines produce iodine from brines.

◀ Chemist Joseph Gay-Lussac named iodine after the Greek word for violet.

▲ Seaweed ash, used in making gunpowder, played a part in iodine's discovery.

◀ Iodine gas.

IODINE AND HEALTH

In 1799, a pharmacopeia was published in London, detailing medical treatments for use by surgeons. Among the recipes and formulations was listed something called the Coventry Remedy for treating bronchocele – also known as a goitre, a distressing swelling in the neck. It had apparently cured a doctor's daughter in the city of Coventry, to great fanfare.

According to the book, treatment must begin the day after the full moon, with a purge. A dose of medicine should then be put under the tongue containing pumice stone, charred cork and burnt sponge, interspersed with bitter powders containing chamomile flower. Patients should be cured within six weeks.

Although it sounds more like a magic spell than a medical therapy, treatment with sea sponge has a long history.

Medieval European doctors knew of sponges' power for treating goitre by the 13th century, and the idea can be traced to China, 4,000 years ago.

▶ Goitre is a swelling in the neck caused by iodine deficiency.

▲ High in iodine, sea sponge was used in Chinese medicine as a treatment for goitre 4,000 years ago.

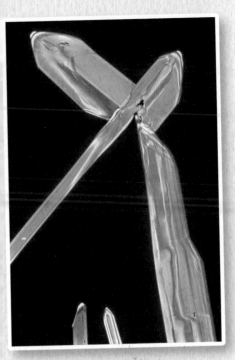

▲ *Crystals of thyroxine, one of the hormones made in the thyroid gland.*

It works because sea sponge contains exceptionally high levels of iodine. The swollen neck of goitre is caused by a deficiency in thyroxine, one of the two thyroid hormones, because of a lack of iodine to make it. The body tries to compensate with high levels of thyroid-stimulating hormone. Other health impacts include abnormal neuronal development before birth – with the risk of intellectual impairment.

IODINE IN TIBET

With several of the world's tallest mountains, Tibet stands over 4,500m (3 miles) above sea level and is often called the roof of the world. Glaciation, rainfall and massive geological processes have fashioned this incredible landscape – and they also have a huge impact on the lives of those who live there.

Over many years, much of the topsoil has washed into lower lands below. What remains is often iodine-poor, as the element leaches readily from the earth. With no other ready source of this vital nutrient, families of subsistence farmers in the area experience life-long health problems – from impaired foetal brain development through to neck goitres in adults.

Unlike a vaccination programme for disease, dietary iodine deficiency requires regular attention. So what was the answer? In other iodine-challenged areas, starting in Switzerland and the USA as early as the 1920s, iodine had been added to table salt, which had hugely reduced the problem of deficiency disorders.

But even though China had also adopted the idea of iodizing salt sold for human consumption, the system was not working well in Tibet. The people there collected their own salt from the high-altitude lakes – and they would barter rather than buy their goods. In the early 2000s, Australian scientists Creswell Eastman and Mu Li worked with the Tibet Department of Health and local leaders to distribute iodized oil capsules to women of child-bearing age, and children under two. Instead of a daily dose, the capsules supplied the required amount for a year.

▲ *A village in Tibet, where iodized salt promoting child development has now become widely accessible.*

Today, iodized salt has become widely accepted in Tibet, reaching 98.7 per cent coverage by 2016. But in the years covered by the iodized oil, hundreds of thousands of children were saved from brain damage.

Fast Facts

ASTATINE At
ATOMIC NUMBER 85

Character: Astatine has many radioactive isotopes. Astatine-210 has the longest half-life of eight hours. It is among the rarest naturally occurring elements in the Earth's crust.

Discovery: Element 85 had a chequered start. Its discovery was first claimed in 1931, when it was called alabamium, and later alabamine, but this proved false. A new claim in 1937 named the element dakin, but this also turned out to be a mistake. Then, in 1936, Romanian Horia Hulubei and Frenchwoman Yvette Cauchois spotted the element via its X-ray spectral lines, in tiny quantities, in the natural decay products of radon. Hulubei suggested the name dor, from a Romanian word relating to peace.

In 1943, Berta Karlik and Traude Bernert found the element in the natural decay chains of uranium isotopes. However, in 1940, a group had synthesized the element using a cyclotron at the University of California, Berkeley.

Name: The Berkeley group benefitted from a proposal in 1947 that the right to name an element should go to those able to reproduce their findings. They therefore suggested the name astatine, from the Greek word *astatos*, meaning unstable.

World sources: At any one time, scientists estimate that the total natural amount of astatine in the Earth's crust is around 30g (1.05oz).

▲ *Berta Karlik.*

▲ *French physicist Yvette Cauchois.*

◄ *A group of scientists including Nobel Prize-winner Ernest Lawrence (far left) discuss building a 184-in (467-cm) cyclotron at the University of California, Berkeley, in 1940.*

THE PROTON MERRY-GO-ROUND

Dale Corson, Alexander MacKenzie and Emilio Segrè made the first artificial sample of astatine in 1940, at the University of California, Berkeley. They bombarded a target of bismuth, element 83, with alpha radiation inside a particle accelerator, and created enough astatine to claim the prize of a new element's discovery.

The machine that had made it possible was the cyclotron, which had been invented only a few years earlier by fellow Berkeley physicist Ernest Lawrence. He was interested in producing particles with enough energy to break up nuclei, and had read about an invention for accelerating particles to incredible energies. It was a linear machine, but Lawrence worked out that he could produce a more compact accelerator that was circular.

Inside his proton merry-go-round, as he called it, the particles would travel in a spiral path guided by magnetic fields, gaining regular speed boosts from an electric field as they whirled.

In 1931, Lawrence and his student Stanley Livingston built a cyclotron that was 11in (28cm) in diameter. In 1932, they had expanded to a 27-in (69-cm) diameter, with larger fields and higher speeds. The rush to increase their machine's speed and energy was because a race was on. In Ernest Rutherford's Cavendish Laboratory at Cambridge

University, John Cockcroft and Ernest Walton had built an accelerator of their own, to a different design. With it, they had been the first to split the atom in 1932, and their work later led to a Nobel Prize.

But cyclotrons were the start of big-budget science in America. Ernest Lawrence built a 37-in (94-cm) cyclotron in 1937, and then a 60-in (152-cm) cyclotron in 1939. At such energies, the beams of particles emerging from the cyclotron were useful for exploring nuclear reactions. They could also enable doctors to create useful radioisotopes for a new field of nuclear medicine.

NEW ELEMENTS

Another application was in colliding particles with carefully chosen targets to make new elements. The 60-in (152-cm) cyclotron was the machine used by Corson, MacKenzie and Segrè to create astatine. It was also the machine that first made neptunium and plutonium.

In 1939, three things happened that impacted the future of accelerator science. First, Ernest Lawrence was awarded the Nobel Prize for Physics for his invention of the cyclotron. Second, he managed to raise funds to build an even larger machine – 184in (467cm) in diameter – with which he hoped to

find a sub-nuclear particle called the pion. But third, the Second World War broke out. All atomic science efforts in Lawrence's lab – and the huge magnet he needed to run his new accelerator – were turned over to the separation of uranium isotopes for weapons.

Fast Facts
TENNESSINE [Ts]
ATOMIC NUMBER 117

Character: Scientists predict tennessine may be a metalloid rather than a non-metal, as the rest of the halogens are. It decays extremely quickly.

Discovery: Produced rather than discovered, tennessine was the result of a collaboration between scientists at the Lawrence Livermore National Laboratory in California, USA, Oak Ridge National Laboratory in Tennessee, USA, and the Joint Institute for Nuclear Research (JINR) in Dubna, Russia, where tennessine was made in 2010.

Name: Tennessine takes its name from the Tennessee region in which Oak Ridge National Laboratory, Vanderbilt University, and the University of Tennessee at Knoxville, are located, all of which contributed to the element's discovery.

World sources: Only a few atoms of tennessine have ever been made.

2
He
Helium

10
Ne
Neon

18
Ar
Argon

36
Kr
Krypton

54
Xe
Xenon

86
Rn
Radon

118
Og
Oganesson

CHAPTER 12

THE NOBLE GASES

WHY WAS GLASSBLOWING VITAL FOR THE DISCOVERY
OF THE NOBLE GASES?

WHAT MAKES XENON A GREAT ANAESTHETIC?

WHO WAS HONOURED IN THE NAMING OF ELEMENT 118?

THE NOBLE GASES

At the far right of the periodic table is a group of calm, collected chemicals called the noble or inert gases. With full electron shells, they do not tend to react with anything. But chemistry is not only about sound and fury. The noble gases include elements that glow, that fuse, and that radioactively decay, making them a fascinating family after all.

For most scientists, discovery of just one element would be an incredible achievement. In the case of the noble gases, Scottish chemist William Ramsay identified the entire elemental group, for which he received the Nobel Prize for Chemistry in 1904. The same year, Ramsay's collaborator Lord Rayleigh was awarded the Nobel Prize for Physics for his co-discovery of argon, as well as investigations of gas densities.

It was finding argon in 1894 that held the key to uncovering this new family of elements. Ramsay and Rayleigh realized that, if argon existed, an entire class of elements could be missing from the periodic table. When Ramsay examined an unknown nonreactive gas produced by heating uranium ore in acid, he thought it might be argon again – but it was helium. This had previously been proposed as abundant in the Sun, but never identified on Earth.

▲ Together with William Ramsay, Lord Rayleigh discovered argon, the first of the noble gases. Lord Rayleigh also gave his name to the phenomenon that makes the sky look blue – Rayleigh scattering, in which blue light, with its shorter wavelength, is scattered more than redder light.

In an 1896 book, Ramsay showed that the relative atomic masses of helium and argon, and their position in the periodic table, hinted that there should be at least three other family members. Through work done with chemist Morris Travers, in 1898 Ramsay went on to isolate three more inert gases: neon, krypton and xenon. Chemist Hugo Erdmann coined the term noble gas for the group, or *Edelgas* in his native language of German.

NEW MEMBERS

The speech marking Ramsay's Nobel Prize in 1904, given by the president of the Royal Swedish Academy of Sciences, called attention to the way the noble gases filled a void between the highly electronegative halogens and the highly electropositive alkali metals. Although Dmitri Mendeleev had been reluctant to include the gases helium and argon in the developing periodic table, he accepted them in 1902 and called them group 0.

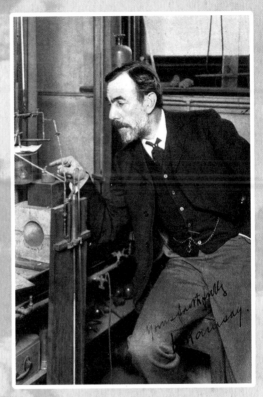

▲ *Scottish chemist William Ramsay co-discovered argon, sharing a Nobel Prize with Lord Rayleigh, and then went on to find helium, neon, krypton and xenon.*

A few years later, in 1910, Ramsay proved that radon was yet another noble gas, produced in the radioactive decay of other elements. And so, with the addition of oganesson in 2006, the family was complete.

Fast Facts
HELIUM He
ATOMIC NUMBER 2

◀ *Helium–oxygen mixtures are used by divers.*

Character: After hydrogen, helium is the most abundant element in the universe. However, because of its small mass, helium is not held by the Earth's gravitational field and so it escapes into space. Our planet produces it very slowly by the radioactive decay of heavy elements such as uranium and thorium.

Discovery: In 1868, astronomers Norman Lockyer in England and Pierre Janssen in France independently realized that there were spectroscopic lines in the Sun's light that did not correspond to any known substance. Lockyer proposed it was a new element, but scepticism remained until 1895 when Per Teodor Cleve and Nils Abraham Langlet in Sweden, and William Ramsay in London, independently found traces of the same substance in an earthly ore of uranium.

Name: Norman Lockyer's name for the new element was helium, after the Greek god of the Sun, Helios.

World sources: Other than for filling balloons, helium is vital for cryogenic applications, including the cooling of superconducting magnets for MRI machines. The USA is the largest helium producer, where the gas is found in natural gas wells, but new deposits have been found, recently in Tanzania.

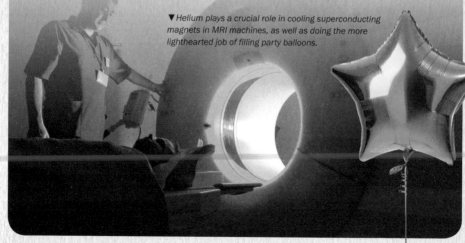

▼ *Helium plays a crucial role in cooling superconducting magnets in MRI machines, as well as doing the more lighthearted job of filling party balloons.*

HELIUM IN THE SUN

▲ In a solar eclipse, the Sun is obscured by the Moon.

Pierre Janssen was an eclipse-chaser of the keenest kind. He had studied at the University of Paris, and went on to a teaching job in the city in 1853. He undertook a research trip to Peru in 1857, which clearly gave him the travel bug, because he found himself crisscrossing the globe on scientific missions for much of the rest of his life.

Solar eclipses intrigued Janssen. Over a period of 38 years, he travelled to see such events from Italy, Spain, Algeria, Thailand, and even the remote Caroline Islands, north of New Guinea. In 1868, he was in Guntur, India, for a famous total solar eclipse in which the Moon blocked all direct sunlight. From a scientist's point of view, this eclipse was significant as the first since Gustav Kirchhoff's spectroscopic theory of 1859. It predicted that the dark absorption lines in the spectrum of sunlight were the characteristic lines of the different elements in the Sun.

▲ French astronomer Pierre Janssen, known by his middle name Jules, discovered helium in the Sun's spectroscopic signature.

◄Pierre Janssen realized that the Sun's prominences were hydrogen gas.

◄An accident in his youth left Janssen unable to attend school, but he succeeded spectacularly in his astronomy career.

SOLAR PROMINENCES

So, what elements would appear in the Sun's spectral signature? Janssen used a spectroscope to observe the solar prominences, eruptions of looping, red-glowing material visible during the eclipse around the Moon's silhouette. He concluded that these were gaseous hydrogen, at incredibly high temperatures. He also spotted a bright yellow spectral line that, while it looked a bit like sodium, was not at the correct wavelength. He realized that this was a new element.

The lines that Janssen saw were so bright that he did not need to wait for eclipses. He could make observations by focussing on the wavelength of light in the prominences, blocking the sunlight scattered by Earth's atmosphere. He wrote to the French Academy of Sciences to say he had discovered helium. Oddly enough, his paper arrived on the same day as observations by English astronomer Joseph Norman Lockyer reaching the same conclusion, and the two are credited with the new element's discovery.

FUSION REACTORS

If you could take a look inside the core of the Sun, you would see atoms of hydrogen fusing together into helium, releasing unthinkable amounts of energy. This reaction provides the heat and light that stream out from every star into space. Even at a distance of 150 million km (93 million miles), it is plenty to power biological life on Earth.

But how do we supply the energy-hungry machines we require for 21st-century living? A clean source of power is the vision of fusion research on Earth – smashing together hydrogen atoms to form helium and getting our hands on some of the energy of a Sun.

For decades, scientific groups across many nations have experimented with fusion. To make atoms fuse, their electrons are first stripped away from the atoms. The resulting plasma must be contained and heated to Sun-league temperatures.

Work in the Soviet Union in the 1950s showed that a doughnut-shaped vessel called a tokamak could contain the plasma within a powerful magnetic field. An alternative approach is called inertial confinement, in which multiple laser beams simultaneously hit a fuel pellet to create the conditions for a fusion reaction.

If you think both processes sound tricky to get right, you would be correct.

▲ The Tokamak nuclear fusion research reactor at the Kurchatov Institute in Moscow.

The trouble so far with fusion has been making it work without putting in more energy than we get out.

The difficulty is overcoming the natural repulsion of the nuclei, which the Sun achieves using gravity and immense pressure, but which, on Earth, requires engineering that scientists are still developing.

HOW FUSION RELEASES ENERGY

Hydrogen isotopes deuterium and tritium are less stable than ordinary hydrogen, and make good candidates for fusion. In the fusion reaction, two unstable atoms become one stable atom of helium, plus a neutron.

The mass of the deuterium and tritium atoms at the start is greater than the mass of the helium atom and neutron at the end. The missing mass is converted to energy as the nuclei fuse together, by Einstein's famous equation $E = MC^2$, where E is energy, M is mass and C is the speed of light.

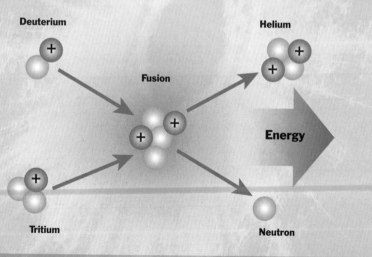

Deuterium

Fusion

Helium

Energy

Tritium

Neutron

SUPERFLUID HELIUM

The isotope helium-4 shows some odd properties. At temperatures just above absolute zero, it flows with zero friction. Instead of staying in its container, it can creep up the walls – or drip straight through the base. It has become a superfluid.

▲ *Danish physicist Lene Hau, whose experiments slow light down by an incredible amount.*

Soviet physicist Pyotr Kapitsa observed this phenomenon, and named it superfluidity in a 1938 paper in the journal *Nature*. Another Soviet scientist, Lev Davidovic Landau, won the Nobel Prize for Physics in 1962 for his work on explaining quantum liquids such as this, and Kapitsa himself received the Nobel Prize for Physics in 1978 for his low-temperature research.

What causes helium-4 to become superfluid? Theory suggests it is because these atoms are bosons. If a gas of bosons is cooled down so that its kinetic energy is reduced enough, the particles drop into the lowest quantum state and coalesce into a Bose-Einstein condensate. It is as if the helium has become a single giant atom.

SLOWING LIGHT

An intriguing property of Bose-Einstein condensates is that they can slow down light. Indeed, in 1998, Danish physicist Lene Hau led a team at Harvard University to slow light from its usual speed of 670,616,629 miles per hour to a paltry 38 miles per hour by using a Bose-Einstein condensate. Experiments since have seen Hau stop light altogether and release it unchanged. Quantum computing and optical data storage may be future applications.

BOSONS AND FERMIONS

Bosons are a type of fundamental particle, including photons and gluons, that make up part of the Standard Model of elementary particles. They behave according to Bose-Einstein statistics, which means they do not mind existing together in the same quantum state. Indian physicist Satyendra Nath Bose conceived the quantum theory behind bosons in 1924/5, and Albert Einstein later developed it with Bose.

◀ Satyendra Nath Bose.

Fermions, another type of particle in the Standard Model, account for the quarks as well as the leptons (which includes electrons and neutrinos). The behaviour of fermions is described by Fermi-Dirac statistics, which shows that these particles cannot occupy the same quantum state. Italian physicist Enrico Fermi and English physicist Paul Dirac discovered the method independently in 1926.

Paul Dirac honoured Bose and Fermi by naming the particles bosons and fermions in a lecture in Paris in 1945.

Albert Einstein. ▶

▼ Enrico Fermi.

Paul Dirac. ▶

GLASSBLOWING AND GAS

Collaborations are vital in science, and can sometimes have scintillating spin-offs. In the case of William Ramsay, a particularly beneficial partnership began years before he joined University College London, where he made his noble gas discoveries.

Ramsay was establishing himself as a chemist in 1881, leading a college in Bristol and carrying out research. In 1882, he appointed an assistant called Sydney Young, and the two of them published more than 30 papers on their work with liquids and vapours.

But it was a practical skill that Young taught Ramsay that proved incredibly useful in the work Ramsay did later. Young had spent a term at the University of Strasbourg where, along with his chemistry research, he learned to blow glass. Young now passed these skills on to his boss, and Ramsay became justifiably proud of the glassware he could make.

The archives at University College London still hold the elaborate glass discharge tubes in which Ramsay collected the elements in the noble gas family, each with its own characteristic colour: yellow for helium, red for neon, red or blue-purple for argon, yellow-green for krypton, and blue for xenon. According to a curator of the collection, the tubes still glow when an electric current is passed through.

▲ *Skills in glassblowing helped William Ramsay identify the noble gases.*

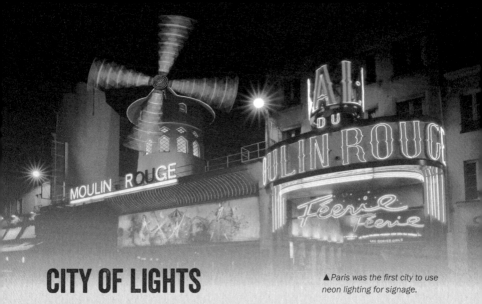

CITY OF LIGHTS

▲ *Paris was the first city to use neon lighting for signage.*

Paris had the first gas street lighting – and it also boasted the first neon lights. In 1910, French engineer and inventor Georges Claude filed his first neon lighting patent. He knew of William Ramsay's work on noble gases, and resourcefully invented a way to liquefy air on a large scale and separate the neon. With some tweaks and experimentation, he managed to make red-glowing neon lighting tubes that were over 6m (19.7ft) long and lasted for 1,200 hours.

Claude unveiled the dazzling light tubes at the Paris Motor Show in December 1910 but, although people loved the spectacle, they were not sure about using them for general illumination. However, for lit signage, the tubes proved ideal. The Cinzano company was soon using the red lights to advertise its vermouth, and the Paris Opera chose them to highlight its entrance.

Bright lights transferred to America when a Californian car dealer bought two of Claude's neon signs after seeing them on a trip to Paris. In 1923, he installed the red neon tubes spelling out the word Packard, with blue borders probably produced by adding mercury to the neon. Known as liquid fire, the glowing signs caused a sensation.

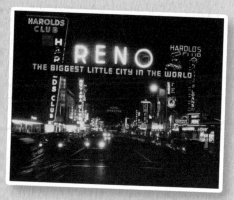

▲ *Bright lights soon spread across the Atlantic, arriving in the USA in 1923.*

GEISSLER TUBES

German physicist and glassblower Heinrich Geissler invented an early kind of gas discharge tube in 1855. He filled his prettily curved glass tubes with low-pressure gases of neon and other kinds, and applied a voltage to make them glow. Geissler intended the tubes as a novelty, but the technology laid the foundation for vacuum tubes that enabled J J Thomson to discover the electron in 1897.

Character: Neon is entirely nonreactive, colourless and odourless. It does not even respond to fluorine.

Discovery: William Ramsay and Morris Travers found neon in their search for members of the noble gas family in 1898. To find it they liquefied air, and used it to liquefy argon. When they separated the argon into its components, they found it contained neon.

Name: Ramsay's 13-year-old son suggested calling the gas novum, from the Latin word for new. Ramsay took him up on the idea, but chose the Greek equivalent *neos*, leading to neon, "the new one".

World sources: Liquid air provides the source of neon, and it is extracted by a process of fractional distillation. The atmosphere contains 6.5 billion tonnes of neon but only a small amount is required commercially.

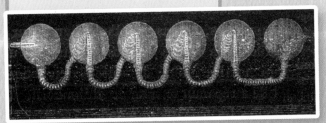

◄ *With a gas such as neon or argon inside, a Geissler tube would fluoresce when an electric current flowed through.*

Fast Facts
ARGON [Ar]
ATOMIC NUMBER 18

Character: Inert and with no biological role, argon is colourless, odourless and tasteless. It can be forced to bond in a complicated chemical process, but on the whole, it is happy by itself.

Discovery: In 1785, chemist Henry Cavendish noted a mystery gas in the atmosphere. Its identification as argon – and the first element in the noble gas family – arose from a great scientific collaboration of the 1890s. Heavyweight scientists John Strutt – Lord Rayleigh – at Cambridge University, and William Ramsay at University College London were intrigued by the fact that nitrogen had different densities, depending on whether it came from the air or by chemical reaction from ammonia. They agreed that Ramsay would look for a heavier gas contaminating the denser sample and that he would seek a lighter gas diluting the less dense. In 1894, Ramsay reacted his sample by burning magnesium in it, producing magnesium nitride plus a nonreactive gas they had not come across before.

Name: Argon gets its unflattering name from the Greek word for idle or lazy. It means "the inactive one".

World sources: Argon forms nearly one hundredth of the volume of the air, and is produced naturally by the decay of potassium-40. Commercially, it is extracted from liquid air and useful for lighting, steelmaking and in double-glazing.

▲ During welding, argon gas pumped over the hot metals helps stop them reacting with the air as they fuse together.

▶ Wine-makers pump argon gas into barrels in order to protect the wine from oxidation.

▲ Physicist John Strutt inherited a title on the death of his father, becoming the third Baron Rayleigh. Today he is known as Lord Rayleigh.

KRYPTON [Kr]

ATOMIC NUMBER 36

Character: Krypton is colourless and odourless like its lighter family members. However, it does form a compound with the incredibly reactive element fluorine: krypton fluoride.

Discovery: In their search for elements in the family of noble gases, William Ramsay and Morris Travers discovered krypton. It forms only one millionth part of the atmosphere, but was present in the argon gas that the researchers had previously extracted from air. When they liquefied the argon and allowed it to evaporate slowly, krypton was left behind.

Name: A French chemist, Marcellin Berthelot, suggested the name eosium in reference to *eos*, the Greek word for the colour of the sky at dawn, but Ramsay preferred the name krypton, meaning "the hidden one".

World sources: A few tonnes of krypton are extracted each year, from liquid air, for industrial use in electric lighting and strobes, as well as laser applications.

▲ *Morris Travers co-discovered krypton with William Ramsay, and then both neon and xenon.*

Krypton glows a soft violet colour in a fluorescent tube, leading to the suggestion that it might be called eosium, from the Greek word for the sky colour at dawn. ▼

XENON [Xe]

ATOMIC NUMBER 54

Character: Xenon produces a beautiful blue glow when in a discharge or vacuum tube, but otherwise is a colourless, odourless gas. One of its rare stable compounds is xenon difluoride.

Discovery: Xenon was another of the noble gases that William Ramsay and Morris Travers found in a sample of argon. They had begun with air, then removed all traces of oxygen, nitrogen and carbon, leaving argon behind. This turned out to contain krypton, neon and xenon in quantities reflecting their abundance in the atmosphere.

Name: Travers and Ramsay thought that a blue-related name would be good for xenon. But they found that all the good Latin and Greek words for blue had been taken by elements and compounds already, so they decided on the Greek word *xenon* meaning "the strange one".

World sources: Extracted commercially from liquid air, xenon is useful in bright lighting, semiconductor applications and research. Xenon also exists in the gases bubbling out of some springs of mineral water.

◄ Car headlamps that use xenon gas are extremely bright, with an icy blue-white colour.

XENON IN ANAESTHESIA

An early public demonstration of surgical anaesthesia took place at Massachusetts General Hospital in 1846. And very public it was, too. The patient, Edward Abbott, breathed in a drug called ether in front of a large audience sitting in steep rows of seats surrounding the bed. The term "operating theatre" had never been more appropriate.

As the patient slept, surgeon John Warren removed a tumour from his neck. Then the patient awoke, unharmed and unaware of any pain, and Warren rightly claimed success. Ether was used as a general anaesthetic until more effective – less flammable – options came along in the 1950s, many of which we still use today. But it was not the only chemical known to dull pain.

▼ *In an early test of general anaesthetics, this patient is mercifully unaware of the surgery being performed – or the onlooking audience.*

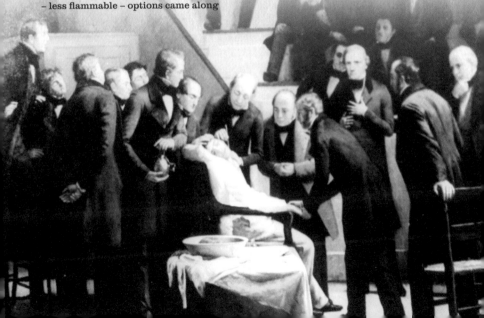

◄ *Modern anaesthetics are administered using equipment that provides the anaesthetic gas and oxygen for breathing, and removes carbon dioxide and other waste gases.*

In an ideal world, anaesthesia would provide many effects alongside numbing pain. It would produce amnesia, muscle relaxation and all with no unwanted impacts – either to health or to the planet. Nitrous oxide has raised environmental concerns similar to ozone-busting CFC chemicals. Other drugs, while in common use, have a number of side-effects in both adults and children.

Nitrous oxide was one of the gases found in 1772 by English chemist Joseph Priestley during his experiments on air. In 1800, Humphry Davy wrote about it while working for the Pneumatic Institute in Bristol, UK. Even though this was a medically inclined organization, Davy's book focussed on the history, chemistry and physiology of the gas – and its recreational uses.

Davy noted that this laughing gas, as nitrous oxide was known, could "destroy physical pain". But, bizarrely, the best use that minds could make of nitrous oxide during the early 1800s was for fun. Laughing gas parties were a huge success among London's elite, and it was not until experiments by dentists in America in the 1840s that its use for pain relief spread.

In 1946, scientists in California found that xenon worked as a general anaesthetic in mice. Human trials proceeded in 1951, and the first human surgeries under xenon took place. It seemed an ideal anaesthetic agent. Patients found it easy to inhale, maintained normal vital signs throughout their procedures, and woke up quickly afterward. No toxic, allergic or carcinogenic reactions emerged. It was also environmentally inert.

Detailed research continues into the effectiveness of xenon. But the main issue with its use is cost. The manufacture of xenon is expensive and energy-hungry, and recycling systems are in their early stages of development.

ANAESTHETIC CHEMICALS

Ether, or diethyl ether as chemists call it, was known as sweet oil of vitriol in the 16th century, when it was made by German doctor and botanist Valerius Cordus. It may have been produced even earlier by Arabian chemist Jabir ibn Hayyan in the 8th century. Its formula is $CH_3-CH_2-O-CH_2-CH_3$.

Nitrous oxide, or laughing gas, is an oxide of nitrogen with the formula N_2O. In 1775, Joseph Priestley described how to produce this gas by heating iron filings dampened with nitric acid.

Xenon, in a mixture 80:20 with oxygen, produced complete anaesthesia with no side-effects in the first two surgeries conducted in 1951, and has been used in many operations since. However, its impact is limited by cost.

▼ *In 19th-century England, nitrous oxide was used for recreational purposes.*

Fast Facts

RADON [Rn]

ATOMIC NUMBER 86

Character: While radon is a colourless, odourless, chemically inert gas like the rest of the noble family, it is radioactive.

Discovery: The picture over radon's discovery is complicated by the fact that it has many isotopes – versions of the same element with different numbers of neutrons in the nucleus. In 1899, Marie and Pierre Curie detected a radioactive gas being given off by radium. From their description of its decay period, this would have been the isotope radon-222. The same year, Ernest Rutherford and Robert Owens noted a gas emanating from thorium samples – this was radon-220, with a half-life of 55 seconds. In 1900, German chemist Frederick Dorn also found radon-222, the same isotope as the Curies, which has a half-life of 3.8 days. Then in 1903, André-Louis Debierne found that actinium also produced radioactive gas, which was radon-219 with an even shorter half-life of 3.9 seconds. Radon today has over 30 known isotopes, all radioactive.

In 1910, William Ramsay and Robert Whytlaw-Gray managed to make enough of the substance to find that it was the heaviest-known gas, and they initially named it nitron, from the Latin for shining, as it seemed to glow in the dark.

Name: Radon was originally the name of the most stable isotope of the new element, radon-222, but is now the elemental name.

World sources: The radioactive decay of radium-226 is one of the main sources of radon gas, and by this route is detected in rocks including granites, gneisses and schists. It is also produced by uranium and thorium as they decay.

◀ *Gneiss.*

▲ *Schist.*

◀ *Granite.*

Fast Facts
OGANESSON [Og]
ATOMIC NUMBER 118

▲ *Particle accelerator equipment at the Joint Institute for Nuclear Research in Dubna, Russia, pictured in 2010.*

Character: Oganesson sits in prime position as the heaviest element currently known. It is highly radioactive, and only a few atoms have ever existed in the laboratory.

Discovery: Controversy surrounded the announcement of element 118's discovery in 1999 at Lawrence Berkeley National Laboratory in the USA, when it turned out to be based on evidence fabricated by a researcher. But then, in 2002, a collaboration between the Joint Institute for Nuclear Research (JINR) in Dubna, Russia, and the Lawrence Livermore National Laboratory announced a sighting that was confirmed in further experiments in 2006.

Name: Like Seaborgium, this element is unusual in having been named after a living scientist. Yuri Oganessian, born in 1933, is considered the world's leading researcher in super-heavy elements. Oganesson received its official name in 2016.

World sources: A period of 80 years elapsed between element 118's prediction by Niels Bohr, and its synthesis. Once made, oganesson's atoms almost instantaneously decay.

▶ *Oganesson is named after Yuri Oganessian, a researcher in super-heavy elements.*

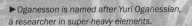

INDEX

ACKNOWLEDGEMENTS

I would like to acknowledge the first-class chemistry brain of Andres Tretiakov, who has given me so much great advice and encouragement. I also want to thank the students at Banbury Academy who came up with amazing science questions, and all whose support has been so vital: Helen, Jan, Richard, Judy, Sara, Robin, and, of course, my family.

PICTURE CREDITS

All illustrations by **Geoff Borin**.

All images **Shutterstock** except: **Alamy** / AF Archive 304c; Age Fotostock 103t; Archive Pics 57b; Archive Pl 225t, 240t, 244bc; Art Collection 94b, 228bl, 283t; Artokoloro Quint Lox Limited 307b; Bilwissedition Ltd. & Co. Kg 195t; Christine Osborne Pictures 253t; Chronicle 207t, 230, 288br; Collection Christophel 299b; DPA Picture Alliance 153br, 187lb, 281t; Everett Collection Inc 108t, 180t; FineArt 258t; Gado Images 219b, 292t; GEOZ 314t; GL Archive 257; Granger Historical Picture Archive 11bl, 181, 283b, 292bg, 308b; Heritage Image Partnership Ltd 33t, 229tl; Itar-Tass News Agency 315b; Interfoto 55r, 282b, 311; Itar-Tass News Agency 301b; James King-Holmes 202b; Keystone Pictures USA 63b, 213b; Lebrecht Music and Arts Photo Library 109; Liam White 16t; LOC Prints and Photographs 227tr; Newscom 11br, 192b; Paul Fearn 53t, 96b, 286, 297, 304t; Peter Horree 205bl; Peter Righteous 122b; Phanie 239t; Phil Degginger 187lc; Pictorial Press Ltd 10b, 59; Prisma Archivo 94r; RGB Ventures / Superstock 12t; Sabena Jane Blackbird 314c; Science History Images 14t, 15t, 31t, 44t, 49, 57t, 65, 87t, 107b, 159, 162, 178r, 183, 214c, 214r, 216l, 217, 248t, 282t, 298, 304bl, 304br; Sciencephotos 77c; Sputnik 158b; Stocktrek Images, Inc. 92b; Susan E. Degginger 268; The Natural History Museum 97t; Universal Images Group North America LLC / Deagostini 84b, 87b; Utcon Collection 269br; World History Archive 64b, 81b, 86, 214l, 276. **Cran Cowan** 103b. **Getty Images** / Bettmann 34–5b, 208t, 246; Christoph Schmidt / DPA 67t; Corbis 150c; DEA / G. Dagli Orti 155t; Digitalglobe/Scapeware3d 77b; Hulton Archives 32, 42–3b, 145tr, 147t, 190b, 300; James L. Amos 155r; Jonathan Nackstrand / AFP 201tr; Joseph Mckeown / Picture Post 188br; Kazuhiro Nogi 196b; Kyodo News 143c; Micheline Pelletier Decaux 23t; Rick Friedman 303; Science & Society Picture Library 41, 42, 50, 51b, 55tl, 55c, 148r, 205br; STR / AFP 22br; Sven Nackstrand 266t; Ullstein Bild Dtl. 52b, 54, 291b; VCG 23b; William Henry Fox Talbot / Hulton 285r; William Stevens 133, 197t © 2014 **Riken**. **Science Photo Library** 20c, 36t, 38t, 72t, 73, 82b, 85b, 101tr, 116t, 149r, 153t, 169r, 174t / A. Barrington Brown, Gonville And Caius College 218t; American Philosophical Society 218b, 281bc, 309, 244bl; Andrew Brookes, National Physical Laboratory 85t; Andrew Lambert Photography 60l, 287; CERN 137; Charles D. Winters 144r; Clive Freeman/Biosym Technologies 209; Dennis Kunkel Microscopy 289l; Emilio Segrè Visual Archives/American Institute of Physics 17tl, 17tc, 17tr, 123t, 158t, 163t, 182, 208br, 231t, 280; Evan Oto 213t; Eye of Science 143t; General Research Division/New York Public Library 101tl; Geological Survey of Canada 151r; GIphotostock 7b, 38bl, 43l, 115b; Hagley Museum and Archive 212t; Herve Conge, Ism 242t; King's College London Archives 216r; Lawrence Migdale 70t; Max Alexander 22tr; National Physical Laboratory (C) Crown Copyright/ Science Photo Library 12b; Power and Syred 243t; Royal Astronomical Society 26; Science Source 271; Sheila Terry 30, 31b, 66, 258b; Sputnik 39, 188c, 232, 315t; Steve Gschmeissner 242b; Universal History Archive/UIG 21tr, 313; Zephyr 171t.